BATTLE ZONE

BATTLE ZONE

FIGHTERS OF THE NATO ALLIES

DOUG RICHARDSON

HAMLYN

First published in 1989 by
The Hamlyn Publishing Group,
a division of The Octopus Publishing Group Ltd
Michelin House, 81 Fulham Road, London SW3 6RB

Series Editor: Len Cacutt
Art Editor: James Hughes
Designer: Brian Folkard
Production: Hilary Stephens

Typeset by MS Filmsetting Limited, Frome, Somerset
Printed in Spain
ISBN 0600 564 363

Contents

Matching Quantity with Quality

Above: **Pale grey paint schemes offer minimal contrast as USAF F-15 Eagles from Bitburg Air Base in West Germany fly over a cloud background typical of European weather conditions.**

From Norway in the north of Europe to Turkey in the southeast, the air forces of the North Atlantic Treaty Organization (NATO) and the Warsaw Pact face one another – the most powerful air arms in history. On one side of what Winston Churchill once called the 'Iron Curtain' are the air forces of the Warsaw Pact, a vast fleet of Soviet-developed (and usually Soviet built) aircraft ranging from the ultra-modern to relics of the mid-1950s. On the other are the air forces of NATO, a numerically smaller force hoping to match quantity with quality, but which sees the latter factor steadily eroded away by newer and more potent Soviet fighter designs.

NATO lacks the monolithic standardization common in the eastern bloc. Each nation has chosen its own aircraft, often out of step time-wise with its neighbor, and often with industrial as well as military goals in mind.

Belgium, Britain, Denmark, France, West Germany, Greece, Italy, Netherlands, Norway, Spain, and Turkey all have national aircraft industries which require a timely flow of orders. Although many NATO warplanes are built on US production lines, industrial consortia stretching across Europe also manufacture aircraft of US or local design.

Take a sophisticated search/intercept radar, two high-thrust turbofan engines, enough fuel to give endurance of over five hours, plus a wing large enough to keep wing-loading low and maneuverability high, integrate all ingredients skillfully, and the result has just got to be the world's 'hottest' figher. That is what the USAF did in the early 1970s with its 'FX' fighter program, and the end result was the McDonnell Douglas F-15 Eagle.

Eagle Gets FAST

To maintain its effectiveness the F-15 has been regularly upgraded throughout its life, and today's F-15C and E models are a far cry from the original F-15A. First modification was the introduction of FAST (Fuel and Sensor Tactical) packs. These were conforml stores which, mounted on the sides of the fuselage, added an extra 9,750 lb (4,420 kg) of fuel for little penalty in terms of drag, and without taking up valuable hardpoints in the way that conventional external tanks do. The F-15C and D models which entered production in 1979 had an upgraded ŕadar able to pick put individual targets from within a close formation.

A key concept of the Eagle program has been a carefully planned Multi-Stage Improve-

Below: By winning the 1975 NATO fighter competition, General Dynamics ensured that the F-16 Fighting Falcon was the most numerically important Western fighter of its generation. By the late 1980s, this high-performance fighter had clocked up more than a decade of NATO service, while its Soviet equivalent, the MiG-35, faced years of further development.

ment Program (MSIP), a concept first applied with great success to the F-16. Aircraft destined for MSIP upgrading are designed to make the installation of new equipment easy. Internal structural cutouts, mounting points and even electrical cabling for future systems may be provided when the aircraft is first built.

MSIP upgrades to the Eagle include an improved APG-70 radar incorporating a programmable signal processor, better EW systems, a more powerful mission computer, long-range HF radio equipment, plus cockpit improvements. These modifications were first introduced on new-build production aircraft. First aircraft to incorporate these changes was F-15C No. 304, delivered in June 1985. Older F-15A, B, C and D aircraft are being modified to the same standard. First up-dated aircraft was

Below: **F-16A fighters of the Royal Netherlands Air Force.**

delivered in 1988 and the entire fleet by 1999. Another improvement program will equip the Eagle with higher-thrust engines, substituting the 29,000 lb (13,150 kg) thrust PW F100-PW-229 Improved Performance Engine (IPE) for the present -100 model.

Until the late 1980s, the aircraft was armed with a mixture of medium-range AIM-7 Sparrow radar-guided missiles and short-range AIM-9 Sidewinder heat-seeking missiles. MSIP has also added facilities for a new weapon – the Raytheon AIM-122A Advanced Medium-Range Air-to-Air Missile (AMRAAM). Sparrow is a semi-active radar missile, and requires that the radar antenna of the launch aircraft remains pointed at the target throughout the duration of the missile flight. As a result, the fighter can engage only one target at a time. AMRAAM is a 'fire-and-forget' missile, flying part of the way to the target under autopilot control, then energizing an active radar seeker for the final attack.

This is an ambitious goal for a missile which is smaller and lighter than the less-sophisticated AIM-7 Sparrow. Development has been protracted, and the first operational rounds were not delivered until late in 1988. F-16 units take priority in the delivery schedule, so it will be a few years before the new missile is in widespread service with the F-15 squadrons.

The USAF's Tactical Air Command (TAC) is now taking delivery of the F-15E, a two-seat model devised for the low-level strike role. First flown in December 1986, this version has an improved Hughes APG-70 radar, the sophisticated LANTIRN pod-mounted nav/attack system, improved EW systems, and about 8,800 lb (4,000 kg) more internal fuel. It can carry up to 23,500 lb (10,680 kg) of ordnance, and retains the full air-to-air capability of the earlier versions.

Three F-15E wings are planned. The first two will have aircraft powered by the PW F100-PW-220 Alternate Fighter Engine, the third will have the 29,000 lb (13,150 kg) thrust IPE engine.

The only problem with 'ultimate' fighters such as the F-15 is their massive price tag. Only the richest of nations can afford to deploy fleets of such costly warplanes, and even the US could not afford to replace older and less expensive aircraft on a one-for-one basis. The USAF currently has more than 420 F-15A and B models, and is buying a total of 469 F-15C and D, plus 392 F-15E.

To create a cheaper aircraft affordable in large numbers to back up the F-15, the USAF held a Lightweight Fighter competition in the early 1970s. Rival designs were the single-engined General Dynamics YF-16 and the twin-engined Northrop YF-17. The GD aircraft was the winner, and was further developed to create the F-16 Fighting Falcon. Virtually a mini-F-15, this has a smaller and shorter-ranged radar, but no long-range radar-guided missiles.

49th Tactical Fighter Wing (USA)

Tactical Air Command (USA)

F-15 pilot's patch (USA)

Potent Little Falcon

It was enthusiastically adopted by the USAF and by the air forces of four NATO nations – Belgium, Denmark, Netherlands and Norway – who set up European production lines for the type. This little fighter proved so potent a package that when the staff of a leading aviation magazine applied fake RAF markings to an F-16 model and 'accidentally' left it lying around the office for a visitor from a rival publication to see, the offering was swallowed hook, line and sinker!

There never was an RAF requirement for an aircraft in the F-16 class, but by the late 1980s a total of 15 export customers had adopted the

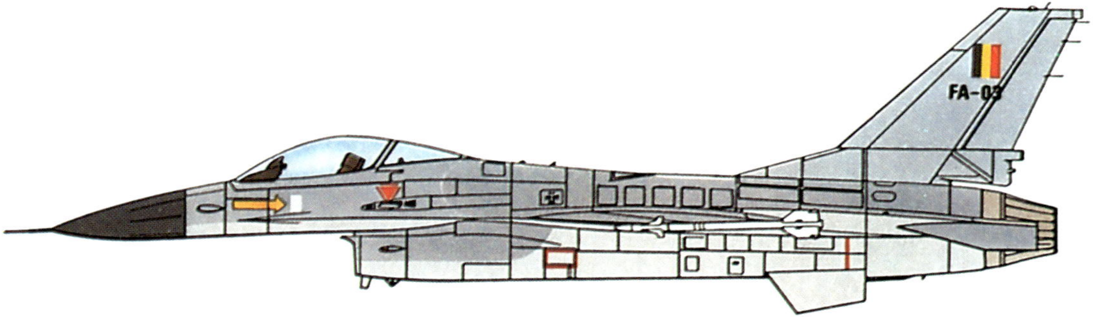

Top: F-16A of the Belgian Air Force's 349 Sqn. This unit was the first Belgian squadron to convert to the Fighting Falcon, and aircraft FA-03 was the first production example built in Belgium.

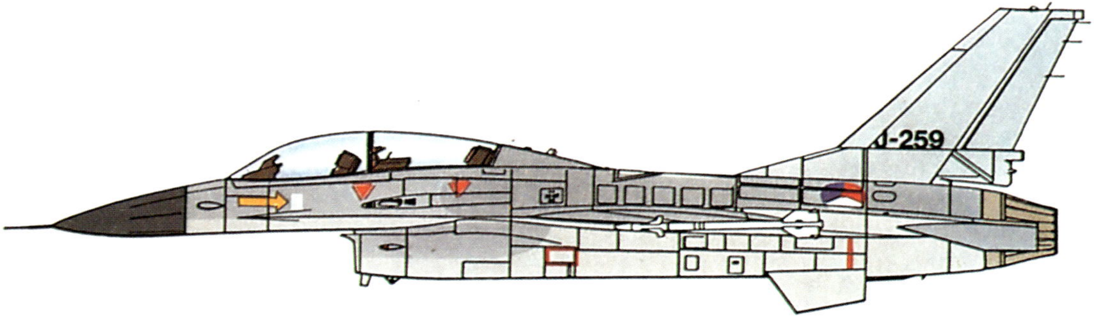

Lower: J-259 was the first F-16A built in the Netherlands by Fokker. It was delivered to 322 Sqn of the RNethAF.

aircraft, including six NATO nations. Delivery of 116 F-16A/B to the Belgian Air Force was completed in 1985, and the type serves with four squadrons – 349, 350, 23, and 31, plus an OCU. A follow-on batch of 44 more was ordered to replace the Mirage 5, with deliveries starting in 1987. An upgrading program is now installing the sophisticated Rapport III EW system, but shortage of funds means that only 56 aircraft will receive the equipment. In Denmark, around 40 F-16A and 10 F-16B equip 723, 726, 727 and 730 Sqns, while a further batch of eight single-seat and four two-seaters was ordered in 1985 for delivery in 1989. The Netherlands ordered a total of 172 F-16A and 27 F-16B, and the type now serves with 322, 323,

Top: Denmark's F-16 fleet is smaller than those of the other NATO air forces which field the Fighting Falcon. This F-16A is from 727 Sqn.

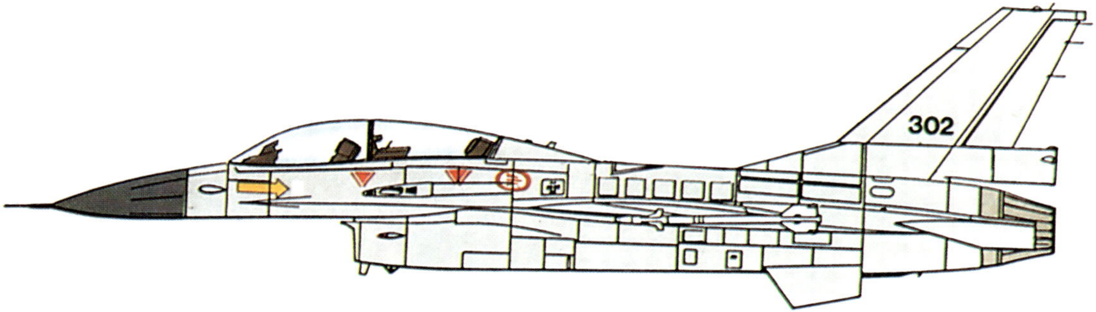

Lower: Norway's Fighting Falcons were the first to have an extended housing at the base of the vertical fin. This houses a braking parachute, and is being retrofitted to RNethAF F-16s to house the Rapport EW system. The aircraft seen here is a Norwegian F-16B two-seat trainer.

311, 312 and 306 Sqns. Deliveries should be completed in 1992. In Norway, 56 F-16A and nine F-16B serve with 331, 334, 332 and 338 Sqns, while a further 24 F-16 are on order for delivery in 1990/91.

Until the mid-1980s, the GD F15 attracted no further NATO orders, losing out to the F-18 first in Canada, then in Spain. When further sales came, they were at the other end of the Mediterranean. Greece decided to split its fighter purchase between the US and France, and deliveries of the first of 40 F-16Cs started in 1988 to replace the F-4. Turkey also adopted the aircraft, ordering 128 F-16C and 32 F-16D. Most will be built in Turkey by the TUSAS consortium, using new custom-built factories. Deliveries from the Murted production line started in 1988. Both nations opted to have their aircraft powered not by the standard F100 engine, but by the more powerful General Electric F110.

Largest NATO operator is of course the USAF, which now has some 735 F-16A and -16B, and plans to purchase a total of 1,936 F-16C and -16D. Like the F-15, the F-16 was designed with evolution in mind, and was targeted for the USAF's first MSIP program.

Externally the resulting F-16C and D models first fielded in 1983 look no different from the older A and B versions, but internal modifications have been drastic. The more capable APG-68 radar has replaced the earlier APG-66, the cockpit has been rebuilt to incorporate wide-range HUD and CRT displays. Later aircraft also have a modified engine bay able to accept either the PW F100 or GE F110. It will also be the first aircraft to carry the AIM-122A AMRAAM missile.

However, the older models are far from obsolete. More than 1,000 older APG-66 sets have already been replaced by the improved APG-68, for example, while some aircraft are receiving a new display processor with increased power and memory.

Tigershark Out, Sparrow In

In October 1986, a modified version of the F-16A was chosen as the USAF's new Air Defense Fighter. This decision brought to an end Northrop's private-venture F-20 Tigershark, and also provided the impetus for installation of the AIM-7 Sparrow, a semi-active radar missile which gives the GD aircraft the ability to engage targets at beyond visual range.

To allow the missile to be used, the aircraft's APG-66A radar must be given an add-on CW illuminator, null filler and tuning antennas which work with the Sparrow guidance system, plus more computer memory. The aircraft also requires a new launch rail (also compatible with the AIM-122A AMRAAM) which mounts on underwing hardpoints 3 and 7, additional electrical wiring, an IFF interrogator, a night-identification searchlight, HF radio, plus up-dated software for the radar, fire-control computer, and stores-

management computer. The braking parachute is deleted.

First Sparrow firing in a series of shots intended to clear the missile for use on the modified F-16A took place on October 18 1988. Modification of the 270 aircraft earmarked for service as Air Defense Fighters is being handled by Ogden Air Logistic Center, at Hill AFB, Utah, with the first rebuilt aircraft being delivered in early 1989. The Sparrow-armed fighters will be flown by 12 squadrons, a mixture of Air Force and Air National Guard units, and will replace the current air-defense F-4 and F-106 interceptors. Although smaller than these aircraft, it has a better range than the F-4, and a better capability against low-signature targets such as cruise missiles.

Egypt, Indonesia, Pakistan, Singapore, Thailand, and Venezuela have all expressed interest in having their early model F-16As (a total of 140 aircraft) reworked to a similar Sparrow-armed standard. Work was also under way in 1988 to add Sparrow missiles to the F-16C. This has been requested by Egypt,

Above: **RNethAF F-16 on patrol with wingtip-mounted AIM-9 Sidewinder missiles. West European Fighting Falcon operators may eventually re-arm their fleets with the longer-range all-weather AIM-7 Sparrow missile.**

This F-16A of the USAF's Okinawa-based 8th Tactical Fighter Wing is an early-model Block 1 production aircraft. The centreline store shown in the lower drawing is a Westinghouse ALQ-119 jamming pod, a unit now replaced by the definitive ALQ-131.

General Dynamics F-16C Fighting Falcon

Role: air combat fighter
Length: 49 ft 3 in (15.01 m)
Height: 16 ft 8.5 in (5.09 m)
Wingspan: 31 ft 0 in (9.45 m)
Weights – empty: 16,794 lb (7,618 kg **Loaded**: 25,071 lb (11,372 kg) **Max. takeoff**: 37,500 lb (17,000 kg)
Powerplant(s): one Pratt & Whitney F100-PW-220 turbofan
Rating: c. 25,000 lb (11,340 kg) in full afterburner
Tactical radius: more than 500 nm (925 km)
Max. range: more than 2,100 nm (3,890 km) with external tanks
Max. speed: greater than Mach 2
Ceiling: 50,000 ft (15,240 m)
Armament: AIM-7 & AIM-9 missiles, one 20 mm cannon and 12,000 lb (5,440 kg) ordnance

which wants to modify 40 of its Block 30 aircraft from late 1991 onwards, and has asked for AIM-7 capability to be provided in some of the Block 40 new-build aircraft due to be delivered from 1991 onwards. The NATO users have yet to request Sparrow armament for their F-16 fleets, but the current Operational Capabilities Upgrade program will make the weapon easy to install.

In 1988 the US asked the four NATO nations which license-build the F-16 – Belgium, Denmark, Netherlands and Norway – to join in two years of pre-development work on an improved version. Known as Agile Falcon, this would have a larger wing of all-composite construction, better avionics and an uprated engine.

The USAF also plans to field a reconnaissance version of the F-16 to replace elderly RF-4C Phantoms. In the short term, the latter are being upgraded by having their traditional film cameras replaced by new electro-optical systems able to give commanders real-time views of tactical targets. This system will be deployed from 1993 onwards, but in the longer term the same equipment will be carried on a centerline pod under the proposed RF-16.

The McDonnell Douglas F-18 Hornet was developed as a spin-off from the USAF's early 1970s Lightweight Fighter competition, and was a development of Northrop's YF-17.

Being a USN aircraft, the Hornet has all the features needed for naval service, but this is no bar to land-based operations. In the late 1970s Northrop offered a specialized F-18L version for land use, but in practise, customers were prepared to forgo the improvements it offered in order to field a warplane which was in US service. Canada was the first NATO export customer, ordering a fleet of 138 CF-18. These now serve with 425 and 441 Sqns and 410 Sqn (the OCU), all in Canada itself, plus 409, 421 and 439 Sqns in West Germany.

Below: Northrop's YF-17 Lightweight Fighter flew in 1974. Rejected by the USAF in favour of General Dynamics' YF-16, it formed the basis of the US Navy's McDonnell Douglas F-18A Hornet. The USN insisted that the Hornet be built by an experienced manufacturer of naval warplanes, a field in which Northrop had no up-to-date expertise.

Early in the aircraft's Canadian service, structural cracks were found, and further deployments were delayed until the vertical stabilizers were modified. Studies have shown that the fatigue life of the aircraft may be less than had been anticipated, so funds have been made available for the purchase of an additional batch of between 12 and 20 aircraft to replace the oldest examples.

The only other NATO customer was Spain, where the EF-18A and B serve with 12 Wing and 15 Wing of the Spanish Air Force. Delivery of all 60 single-seaters and 12 two-seat trainers is due to end in 1990. Suggestions that France might buy 30 Hornets as replacements for the F-8FN Crusader and Super Etendard which currently fly from French aircraft carriers were denied in late 1986 by French Defense Minister Andre Giraud.

Latest versions of the Hornet to be delivered are the single-seat F-18C and two-seat F-18D. Externally similar to the earlier versions, these are fitted with the Martin-Baker NACES ejection seat, an improved mission computer with higher speed and a greater memory capacity, the Advanced Self-Protection Jammer (ASPJ), plus a flight-incident data recorder/monitoring system able automatically to adjust the aircraft cg by controlling the fuel flow from the various tanks. It also has provision for AGM-65D Maverick IR guided ASMs missiles, the planned ASRAAM AIM-132 dogfight missile, and for add-on reconnaissance equipment.

Free of missiles or other underwing stores, a Spanish Air Force EF-18A displays its clean lines during a vertical climb. Note how the engine nozzles are not fully open – the lightly-loaded aircraft does not need full afterburner to fly this manoeuvre.

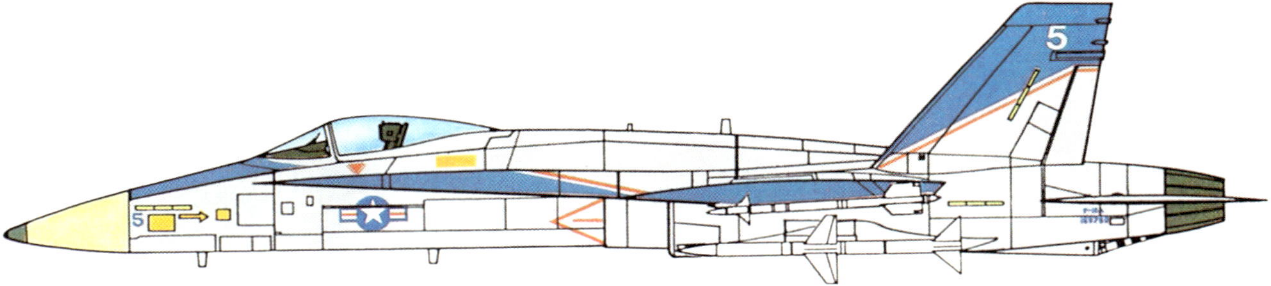
5
5

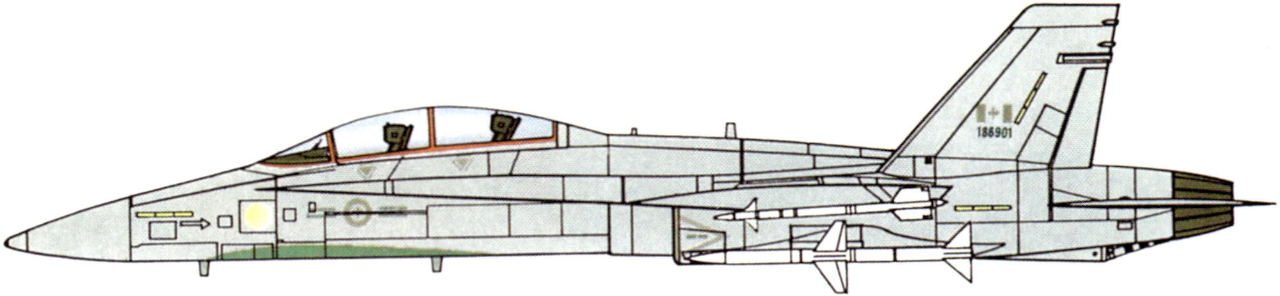
188901

Right: Afterburner nozzles fully closed, this Spanish EF-18 cruises at medium altitude using the dry thrust of its twin GE F404 engines.

Opposite top: when first flown, the fifth F-18A prototype had the original pattern of wing leading edge, which incorporated a dogtooth. This was eliminated on production aircraft as part of a series of modifications aimed at improving the roll rate.

Opposite centre: this Canadian two-seat CF-18 shows the extended canopy of the trainer version, also the Canadian low-visibility markings and the fake canopy painted on the underside of the forward fuselage.

Opposite bottom: Spain is the only West European nation to have ordered the F-18.

First flight of the F-18C was on 15 September 1986. So far the type is only in US service, but its features are bound to make the Hornet more attractive on the export market.

McDonnell Douglas has proposed a series of further-improved Hornet variants under the general designation 'Hornet 2000'. These range from an F-18C/D with improved avionics, through a model with uprated engines plus extra fuel housed in a dorsal tank, and another model which would add a wing of extended chord plus a new tailplane, to a radical redesign using a new cranked delta wing, canards and control-configured vehicle (CCV) technology. So far, however, there have been no takers, and the future of the project seems less assured than that of the rival 'Agile Falcon'.

In the 1970s, McDonnell Douglas published in European aviation/defense journals an advertisement which invited the reader to look out of the nearest window. If it wasn't dark or raining, the advert stated, it soon would be. It was a powerful summary of the arguments in favor of one of the great warplanes of the era – the F-4 Phantom II.

That 'II' is normally left out whenever the designation is printed – talk to all but the most hardened aviation enthusiast about the McDonnell Douglas Phantom, and it's not likely he will think you mean the company's late-1940s FH-1 Phantom

Big, Brutal and Ugly

Big, brutal and ugly, the Phantom II looked like a design gone wrong when first unveiled in early 1958. It was a design to disprove the old adage 'If it looks right, it'll be right', and also one which would cause a fundamental re-think in fighter technology. Instead of being a traditional fast-climbing, agile, gun-armed dog-fighter of the sort that fighter jockeys dream about, this was a massive beast which offered near-zilch rear view, needed a back-seat operator to fuss over its complex radar, and had traded guns for air-to-air missiles. Yet beneath the skin it was still a fighter, combining fast acceleration and high top speed with a climb performance which shattered records in near-effortless manner. Although never a forgiving aircraft, the Phantom was the best fighter of its generation.

Below: **West Germany originally considered buying a simplified single-seat model of the F-4 Phantom, but finally opted for the two-seat F-4F.**

Based on USN carriers, and adopted by the USAF for land use, it saw extensive combat during the Vietnam War, enabling the US to maintain its traditional air superiority deep into enemy skies. Just how good it was did not emerge until the early 1980s, when performance curves released by Boeing as part of its sales campaign for a proposed F-4 re-engining scheme showed how even the standard J79-powered aircraft could outfly the MiG-21 and MiG-23.

Despite the Phantom's age, and the growing numbers of F-15s and F-16s, the aircraft is still one of NATO's most important front-line fighters. The USAF is still the biggest user, with more than 1,200 F-4C, D, E, F and G serving with the regular air force and the Air National Guard, alongside a fleet of more than 300 RF-4C reconnaissance aircraft.

First NATO export customer was the UK, which ordered a total of 170. This was made up of 52 F-4K for the Royal Navy, and 118 F-4M for the Royal Air Force. These were powered not by the standard GE J79 turbojet, but with a custom-designed afterburning version of the Rolls-Royce Spey turbofan. The new engine dictated a redesign of the intakes and the rear fuselage, but the resulting cost and performance problems led the aircraft to be dubbed the 'world's slowest and most expensive Phantoms', and effectively ruled out any follow-on order.

Due to the then Labour Government's policy

The huge jet pipes of this RAF Phantom house Rolls-Royce Spey turbofans. This give more thrust than the standard J79 engine, but the new power plants proved troublesome when first installed.

F-4F of the West German Luftwaffe's JG74 'Molders'. A late 1980s modification programme is upgrading these aircraft with new radar and the AMRAAM missile.

McDonnell Douglas F-4E Phantom II	
Role: multi-role fighter **Length**: 63 ft 0 in (19.2 m) **Height**: 16 ft 3 in (4.96 m) **Wingspan**: 38 ft 5 in (11.70 m) **Weights – empty**: 29,535 lb (13,397 kg) **Max. takeoff**: 61,795 lb (28,030 kg) **Powerplant(s)**: two General Electric J79-GE-17 turbojets	**Ceiling**: 56,120 ft (17,100 m) **Armament**: 16,000 lb (7,250 kg) of ordnance, plus one 20 mm M61 cannon **Rating**: 11,870 lb (5,384 kg) dry thrust, 17,900 lb (8,119 kg) with after burner **Tactical radius**: 700 nm (1,295 km) combat air patrol with external fuel **Max. range**: c. 2,000 nm (3,700 km) **Max. speed**: Mach 2.2

Another upgrade scheme is improving the Luftwaffe's RF-4E reconnaissance aircraft, giving them the ability to carry strike ordanance.

Right: the F-4 Phantom could never be described as 'beautiful', and the fuselage of the RAF version is more portly than other Phantoms due to the use of Spey engines.

of running down armed forces, only one of the Royal Navy's carriers ever deployed the Phantom, and when that vessel was withdrawn from service in 1978 the surviving F-4K were passed over to the RAF.

More than 100 Spey-powered Phantoms still serve with the RAF, equipping 43, 56, and 111 Sqns in the UK, plus the UK-based OCU 228 Sqn. Two squadrons are based in West Germany (19 and 92), while 23 Sqn operates from the Falkland Islands. To make aircraft available for the South Atlantic, the depleted fleet had to be boosted in strength by the purchase of ex-USN F-14J aircraft, 14 of which equip the UK-based 74 Sqn.

West Germany was the next NATO customer. It initially purchased 88 RF-4E, and was sufficiently pleased with them to order 175 F-4F fighters. At one time, these were planned as a new single-seat version of the aircraft, but more sensible councils prevailed and the F-4F emerged as a two-seater.

Below: Until the arrival of its first F-16s, the F-4E was the most important combat type in Turkish service.

Maverick Phantoms

At present, 74 F-4F are tasked as interceptors, another 80 as strike aircraft. To suit the aircraft to these rules, two distinct modification schemes were devised. Completed in 1983, 'Peace Rhine' gave the strike F-4Fs the ability to carry and fire AGM-65 Maverick ASMs while the interceptor upgrade involved the installation of the Hughes APG-65 radar (the type used on the F-18 Hornet), new IFF and AIM-120A AMRAAM fire-and-forget missiles.

The Luftwaffe still has 79 of its RF-4E recce aircraft, which now carry improved sensors (better cameras, plus an IR linescan sensor). Chaff/flare dispensers have also been added, and the RF-4E have also been modified to allow their use as strike aircraft. At batch of 10 F-4E was also purchased. Eight are still in service, and are based in the US for training purposes along with 40 Luftwaffe T-38A Talon trainers.

Spain was the next NATO nation to get Phantoms. In April 1971 it received 36 ex-USAF F-4C, followed by four RF-4C. Greece received 56 F-4E ex-USAF F-4C, followed by four RF-4C. Greece received 56 F-4E and eight RF-4E in 1974, and still operates 49 and seven respectively. Later the same year Turkey took delivery of its first F-4E, and more than 90 are now operational, alongside eight RF-4E.

Most exotic Phantoms in NATO service are the USAF's F-4G Wild Weasels. Built in the mid-1970s by converting 116 F-4E, this carries a complex suite of avionics for the detection and attacking of hostile radars. Heart of this is the complex APR-38 sensor system, which detects and classifies hostile emitters, allowing the crew to carry out attacks using AGM-45 Shrike or AGM-88 Harm anti-radar missiles.

The F-4G Wild Weasel fleet is currently being improved by the installation of a high-speed computer – known as the Weasel Attack Signal Processor (WASP) – in their radar-hunting APR-38 sensor suite. This has eight times the memory and seven times the speed of the unit it replaces. More complex tactical situations can now be handled, and anti-radar ordnance can be delivered on target more quickly. Deliveries of new processors to the 37th TFW at George AFB, California, started in the autumn of 1988.

Working on the second half of the upgrade, E-Systems found itself faced with demanding environmental and packaging restrictions. The company delivered a prototype receiver in March 1987, but this seemed likely to need a major redesign which would have stretched the program out by two more years. Instead TAC opted to cancel the contract, and to plan for a replacement Wild Weasel aircraft.

Eagle's Edge Over Wild Weasel

Rival designs competing for the Follow-on Wild Weasel (FOWW) contract are the F-15, F-16, Tornado, the ATF, the A-12 or even a reworked F-4G. Selection of the winning candidate is expected in the autumn of 1989, but given that the Eagle is already in USAF service in the strike role the McDonnell aircraft seems to have the edge over its rivals.

Good export sales but limited NATO acceptance were the hallmarks of another of the great warplanes of the late 1960s and early 1970s – the Dassault-Breguet Mirage III. The original Mirage III-01 prototype, flown for the first time at the height of the 1956 Suez crisis, was a promising design but was essentially a light delta-winged interceptor in the MiG-21 performance class. When SNECMA offered a new model of Atar engine, the 9K-50 turbojet, offering a 33 per cent increase in maximum thrust, the Dassault team reworked the design, adding the new engine, and a revised wing of greater area and thinner profile. A revamped fuselage also gave the internal volume needed for modern avionics, including a nose radar. Designated Mirage IIIA, this interim version was Western Europe's first Mach 2 design, and proved the technology which would be used in the definitive Mirage IIIC production design.

In the late 1960s and early 1970s, the Mirage IIIC was France's main front-line fighter. It was soon joined by the slightly longer and heavier Mirage IIIE, a multi-purpose fighter-bomber. Most famous of the early export users was Israel, which used the III CJ to spearhead the air attacks which opened the 1967 Six Day War. France tried hard to have the aircraft accepted by NATO as its standard fighter. Sole NATO ally other than France to operate the Mirage III was Spain, which still has 18 of the 30 Mirage IIIE it purchased, plus 18 of the 25 IIID trainers.

Dassault had slightly more success on the NATO front with the Mirage 5 fighter-bomber.

Below: Dassault's Mirage IIIC entered French Air Force service in the early 1960s. A sale to Israel paved the way for massive export success, but the aircraft was largely ignored by NATO.

Originally developed for Israel, this is a daylight-only ground-attack aircraft which trades avionics for extra fuel and ordnance. Israel's 50 Mirage 5J were never delivered to their original customer, which after the Six Day war found itself at the receiving end of a French arms embargo. The aircraft were redesignated Mirage 5F, and taken into French service.

The Mirage 5 proved to be just what Belgium needed in the late 1960s to replace its obsolete F-84s, and to create production work for its aircraft industry. By the time that local production ended, the Belgian Air Force had taken delivery of 106 single-seaters and 78 two-seaters. More than 30 single-seat Mirage 5BA still serve with 1 and 2 Sqns, along with the 17 Mirage 5BR reconnaissance aircraft of 42 Sqn, and the ten two-seat Mirage 5BD trainers of

Dassault-Breguet Mirage IIIE

Role: multi-role fighter
Length: 49 ft 3.5 in (15.03 m)
Height: 14 ft 9 in (4.50 m)
Wingspan: 6 ft 11.5 in (8.22 m)
Weights – empty: 15,540 lb (7,050 kg) **Loaded:** 21,165 lb (9,600 kg) **Max. takeoff:** 30,200 lb (13,700 kg)
Powerplant(s): 1 × Snecma Atar 9C + 1 × SEPR 844 rocket
Rating: 13,227 lb (6,000 kg) with A/B:3,300 lb (1,500 kg)
Tactical radius: 647 nm (1,200 km)
Max. speed: Mach 2.2
Ceiling: 55,800 ft (17,000 m)
Armament: 8,810 lb (4,000 kg) of ordnance plus two 30 mm DEFA cannon

Above: As newer Mirages entered service, the Mirage IIIC was assigned to second-line roles. This example served in Djibouti in East Africa with EC 3/10 'Vexin'.

Opposite: the highly-successful Mirage F1 combined technology proven in the delta-winged Mirage III and 5 with better avionics plus a swept wing. Some are equipped for in-flight refuelling, allowing them to fly long distances to support French military commitments in areas such as Africa.

Left: following the collapse of the over-ambitious ACF programme in the early 1970s, Dassault combined the proven delta configuration with a new engine and a fly-by-wire control system to create the Mirage 2000.

Above: **Greece operates the Mirage IIICG, and was sufficiently impressed with the aircraft that it split its order for next-generation fighters between Dassault and General Dynamics.**

8 Sqn, the Mirage OCU. Final retirement of Belgium's Mirage 5 force is not due until 2005, so the remaining aircraft ware being given a limited avionics and structural upgrade, plus new ejector seats.

With the arrival of the newer Mirage F1.C and the Mirage 2000, the French Air Force was able to run down its Mirage IIIC force, and only a handful are now operational. The two-seat Mirage IIIB trainer remains an important type with almost 40 in service, training aircrew assigned to the 120-string Mirage IIIE force and to the remaining 31 Mirage IIIR and IIIRD reconnaissance aircraft and 41 Mirage 5F.

Dassault, (since December 1971 Dassault-Breguet) has continued to develop the Mirage III/5 series, offering the Mirage 50 series. Powered by the 15,800 lb SNECMA Atar 9K50 turbojet from the later Mirage F1, this was marketed from 1975 onwards. Although it won some export orders it was never as successful as the earlier models, and no NATO orders were placed. The even more advanced Mirage IIING introduced canard control surfaces, a fly-by-wire system, plus the latest generation of avionics, including HUD. A single prototype was flown in 1982, but attracted no takers. Like the F-5E, the Mirage 50 and IIING shared the market with the more exotic F-16, and lived in the shadow of the GD warplane.

To extend the life of existing delta Mirages, Dasault-Breguet now offers update schemes which add to the aircraft an inertial platform, laser rangefinder, digital computer and HUD, but this is unlikely to be applied to French, Belgian, or Spanish aircraft.

The limited success of the Mirage III/5 series within NATO must have been a disappointment to Dassault, but it is not a reflection on the aircraft's qualitites. By the time that production ended in the early 1980s, the list of non-NATO customers was stunning – Agrentina, Australia, Brazil, Chile, Colombia, Egypt, Gabon, Israel, Lebanon, Libya, Pakistan, Peru, South Africa, Switzerland, UAE, Venezuela and Zaire. With a commercial success of this magnitude on his hands, Dassault must have cried all the way to the bank.

With the swept-wing Mirage F1, history repeated itself for Dassault, with limited NATO success outside of the home market, but a foreign order book beyond NATO which included Ecuador, Iraq, Jordan, Kuwait, Libya, Morocco, Qatar, and South Africa.

France is the main NATO operator, and the only nation in the world whose air fleet numbers more than 100. Most are the basic Mirage F1C interceptor model and can be converted to the F1C-200 standard by the addition of a detachable flight-refuelling probe. A total of 145 are in service, along with 19 two-seat F1B trainers.

Like all versions of the Mirage F1, the F1C is powered by a single SNECMA Atar 9K50 turbojet. An improved version of the engine used in the Mirage III and 5, this develops 15,870 lb of thrust in full afterburner. In the fighter role, it can carry two R.530 Magic, Supper 530, or AIM-9 Sidewinder air-to-air missiles, while its two internally mounted 30 mm DEFA cannon each have 125 rounds of ammunition.

Final French Air Force version is the F1CR-200 reconnaissance aircraft, which carries an internal sensor fit of Omera cameras, a SCM2400 IR sensor, and a Thomson-CSF Raphael sideways-looking radar. The reconnaissance model also has a SAGEM Uliss 47 inertial navigation system.

Having lost out to the US in the late 1950s when the F-104G Starfighter was selected by West Germany, Dassault was determined to win the 1975 competition to select a new fighter for four NATO air forces. He offered the Mirage F1E, a multirole version of the basic F1 design. This was powered by the new Snecma M53 single-shaft turbofan. The new engine gave the aircraft a better acceleration, plus the ability (in theory at least) to reach speeds of up to Mach 2.5. In practise, the canopy would probably have required re-design to cope with the increased thermal stress.

'Pedestrian' Mirage?

Despite these improvements, the aircraft was somewhat pedestrian when compared with the US offerings – the General Dynamics YF-16 and Northrop YF-17. At the 1975 Paris Air show, Dassault faced the embarrassment of hearing that the GD aircraft had been selected. Hurriedly removed from the flying display, the F1E was swiftly repainted in a bold red/white/blue color scheme, then brought back to the show to resume its flying displays. Such Gallic bravado could not conceal the fact

EdA Ala de Caza 14, Esc 14. (Spain)

EC 1/12 (France)

that the aircraft had suffered a major sales defeat, and would now see only limited NATO service.

The Greek Air Force's Mirage F-1CG is broadly similar to the French interceptor, and in the early 1980s remained Greece's most potent fighter. Forty were ordered, and 34 are still in service. As a Mirage III operator, Spain was an obvious market for the newer aircraft. Its order for 72 aircraft was split between three models – the F1C interceptor, F1B trainer and the F1E multirole model. To date, nine aircraft have been lost, and the Spanish force now numbers 39 F1C, 19 F1E and five F1B.

The F1E retains the nose radar and twin-cannon armament of the interceptor, but like the radar-less F1A ground-attack version it has five hardpoints for ordnance. It can carry up to 8,818 lb (4.000 kg) of stores, such as eight 1,000 lb (450 kg) bombs, or four 36-round rocket launchers, or one AS.30 or AS.37 Martel ASM. Iraq's F1EQ6 version is fitted with Aero-

Trainer version of the Mirage F1 series is the two-seat F-1B. This example has been painted in Middle East camouflage, but the insignia of the operator (in this case probably Jordan) will not be applied until delivery.

spatiale AM.39 Exocet anti-ship missiles, the weapon which badly damaged the US Navy frigate *Stark* during an incident in the Persian Gulf.

Unlike the 'F1E' offered to NATO in 1975, this aircraft retains the Atar turbojet. Despite the performance advantages promised by the M53 turbofan, the company never re-offered the newer engine to potential customers. It will be interesting to see if it is ever offered as a mid-life update for the F1 series – still a relatively young aircraft with a lot of flying hours left.

Delta-winged aircraft are not the easiest types for an inexperienced pilot to fly, particularly on the approach, when a relatively high speed is combined with a nose-high attitude.

Dassault-Breguet Mirage F.1C

Role: interceptor
Length: 50 ft 2.5 in (15.3 m)
Height: 14 ft 9 in (4.5 m)
Wingspan: 27 ft 6.75 in (8.40 m)
Powerplant(s): one Snecma Atar 9K-50
Rating: 15,873 lb (7,200 kg) in full afterburn
Max. speed: Mach 2.2
Ceiling: 65,600 ft (20,000 m)
Armament: 13,900 lb (6,300 kg) or ordnance plus two 30 mm DEFA cannon

Top: this side view of a Greek F1.BH shows the addition of a second seat barely mars the attractive lines of the Mirage F1.

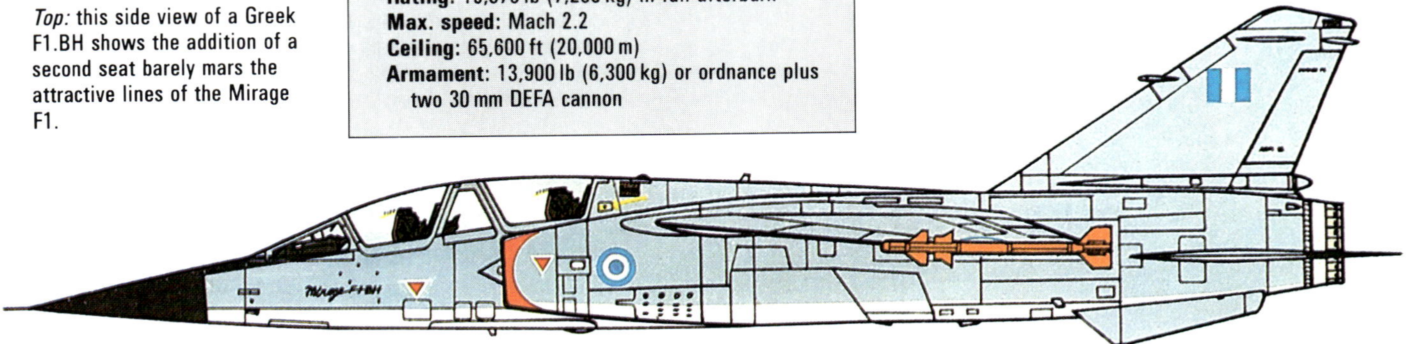

Lower: the Mirage F1.CR is France's main tactical reconnaissance aircraft. Note the refuelling probe, and the fairing under the nose for optical and IR sensors. A centreline pod for elint has now been developed for these aircraft.

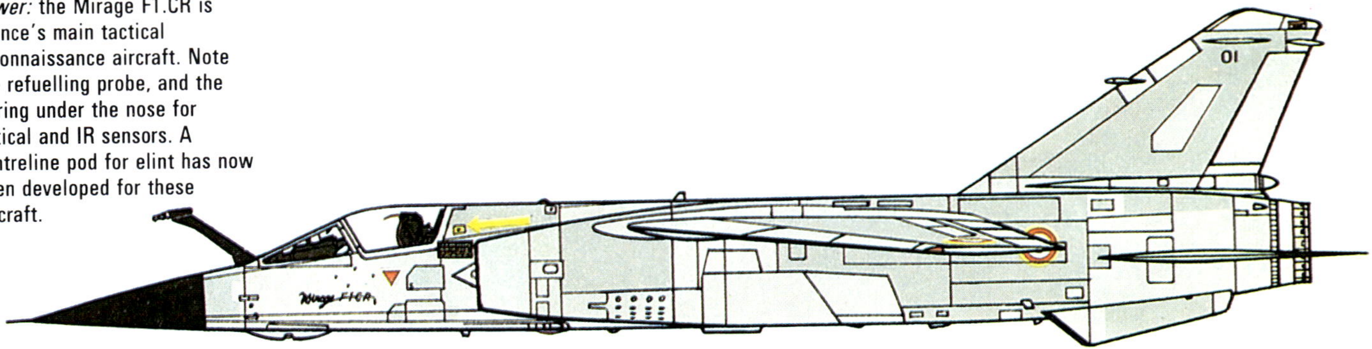

The swept wing of the F1 allows lower approach and touchdown speeds, and enables heavily loaded aircraft to take off using runs 20 per cent shorter than would be the case in a delta Mirage. On a typical interception mission, take-off and landing runs can be as short as 2,100 ft (640 m).

The Mirage F1 also carries some 265 gallons (1,000 l) more fuel than the Mirage III. This results in a 'longer-legged' fighter with three times the high-altitude patrol endurance and supersonic dash endurance, while tactical radius is doubled at the low level typical of modern combat operations.

Had the Mirage 2000 been available in 1975, Dassault would have stood a much better chance of defeating the YF-16. Unfortunately for France, the project was launched too late and thus seems set to do poorly in the export stakes. It was to be the Mirage F1 story re-enacted, with France being the major customer for the aircraft, while the F-16 and F-18 took most of the NATO orders. The US warplanes also did well on the export market outside of NATO, and the steady sales of the F-16 ensured that the new Mirage would not be able to match the sales of the F1, let alone the Mirage III/5 series.

The French Air Force formed its first operational Mirage 2000C squadron in 1984, but had to make do with a compromise radar. The aircraft was planned to carry a Thomson-CSF RDI radar optimised for the French interception mission, but it soon became obvious that this would not be ready in time. Early aircraft were delivered with the lower-performance RDM set, a non-coherent multimode radar originally planned for use of export examples. In practise, all export aircraft have received the higher-performance RDI and early French Air Force 200Cs are being retrofitted. A total of 323 Mirage 200s is planned.

Designations of the Mirage 2000 family partly follow the traditional French pattern. The French Air Force operates the 2000C interceptor, 2000B two-seat trainer, but did not buy the 2000R reconnaissance model. A 2000E multirole version might reasonably have been anticipated but the next French version was in fact the 2000N two-seat strike aircraft. In this case 'N' stands for nuclear. The aircraft is armed with Aerospatiale's nuclear-tipped ASMP medium-range supersonic cruise ASM. Deliveries started in 1988, and the force will eventually number 70 aircraft.

Spin-off from the nuclear-armed 2000N is the two-seat 2000N-1. Originally reported as the 2000P, this will carry a conventional rather than nuclear strike payload. A force of up to 32 is planned. Most recent export version is the 2000S, a strike version first offered to Kuwait in 1988.

Sole export customer within the NATO alliance is Greece, which ordered 36 multi-role 2000E, plus four 2000B trainers. The logic behind splitting Greece's 1980s fighter buy

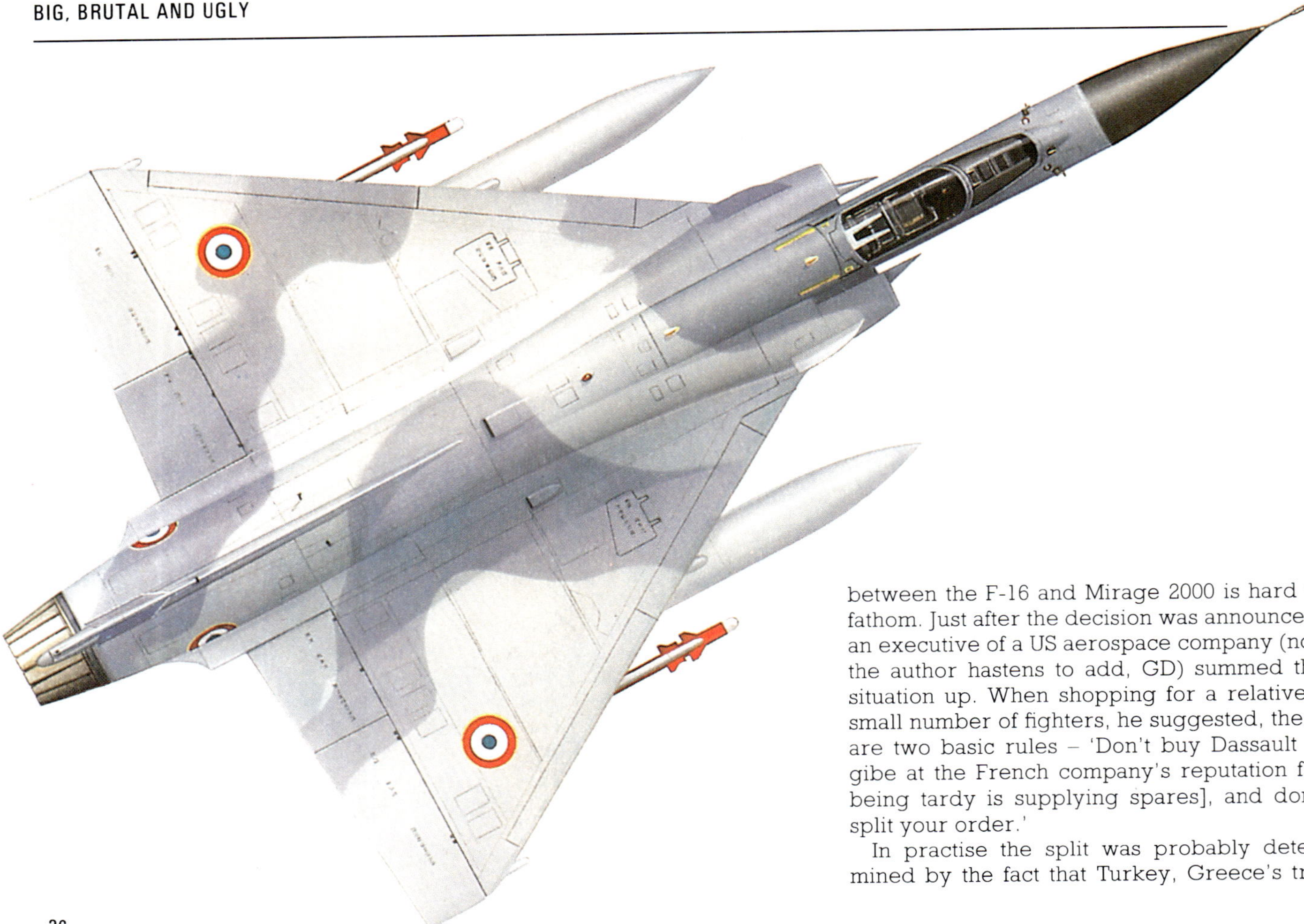

between the F-16 and Mirage 2000 is hard to fathom. Just after the decision was announced, an executive of a US aerospace company (not, the author hastens to add, GD) summed the situation up. When shopping for a relatively small number of fighters, he suggested, there are two basic rules – 'Don't buy Dassault [a gibe at the French company's reputation for being tardy is supplying spares], and don't split your order.'

In practise the split was probably determined by the fact that Turkey, Greece's tra-

Dassault-Breguet Mirage 2000C

Role: air-superiority fighter
Length: 47 ft 1.25 in (14.36 m)
Height: 17 ft 0.75 in (5.20 m)
Wingspan: 29 ft 11.5 in (9.13 m) **Weights – empty:** 16,534 lb (7,500 kg) **Loaded:** 23,940 lb (10,860 kg) **Max takeoff:** 37,480 lb (17,000 kg)
Powerplant(s): one Snecma M53-P2 turbofan
Rating: 14,462 lb (6,560 kg) dry thrust, 21,385 lb (9,700 kg) with A/B
Tactical radius: greater than 400 nm (740 km)
Max. range: 1,800 nm (3,335 km) with external fuel
Max. speed: Mach 2.2
Ceiling: 59,000 ft (18,000 m)
Armament: 13,890 lb (6,300 kg) of ordnance, plus two 30 mm DEFA cannon

In the years before the Second World War, France produced some of the world's ugliest aircraft, but post-war Dassault designs have been among the most beautiful. The Mirage 2000 shown here combines good looks with high performance, plus a freedom from US Government controls over export deals. When selling warplanes, Paris takes a more liberal view of potential clients than does Washington.

ditional enemy, also flies the F-16. A two-aircraft fleet ensures that although Greece now has what is widely regarded as the West's best light fighter, Turkey would also be faced in any Aegean conflict with a Greek fighter which its pilots were not familiar. Non-NATO sales have been to Abu Dhabi, Egypt, India, Jordan and Peru, but the numbers involved have been small.

Ten years ago, one of the most important strike aircraft based in Europe was the Lockheed F-104 Starfighter. In service with seven nations, most of whose fleets were numbered in three figures, this needle-nosed fighter had been the largest European construction program of its era, involving five production lines in four countries. At one time the West German fleet numbered about 700, but the last were retired from service in October 1987 when Luftwaffe unit JBG34 switched to Tornado. Of that once-proud (and certainly controversial) German fleet, some 30 now serve in general 'support duties', a mixture of single-seat F-104G and two-seat TF-104G.

Hot and Unforgiving Starfighter

In its day, the Starfighter acquired a reputation of being a 'hot' ship. Developed following the Korean War, it flew for the first time in 1954, and entered service with the USAF in 1958. The basic F-104A, and the follow-on F-104C with uprated J79-GE-7 engine, weighed around 20,000 lb (9,000 kg) fully loaded. Designed for all-out speed performance, it had minimal avionics and limited success.

In the mid-1950s, Lockheed redesigned the aircraft, cramming in half a ton of avionics, an uprated J79-GE-11A engine and the ability to carry warloads of two on even three tons. The resulting F-104G was offered to NATO in 1958 to meet what was essentialy a West German requirement. At that time the Luftwaffe was operating the F-86 Sabre, and was determined to get into the Mach 2 field. When the Germans selected the F-104G in February 1959 they bit off more than they could chew.

With the air forces of Belgium, Canada, Denmark, Italy, Netherlands and Norway, the F-104G had a fairly successful career but for the generally inexperienced Luftwaffe of the 1960s it was a near disaster. By the mid-1960s, on average one Luftwaffe F-104G was crashing every 10 days. The Starfighter had a reliable engine, was strongly built, and had what was for the period an advanced and accurate nav/attack system. But its tiny wing had been calculated for much lighter operating weights, so careful piloting was required. The F-104G was unforgiving and also made severe demands on the Luftwaffe's then-limited maintenance skills. Eventually it was made to work, giving West Germany the experience and know-how to tackle its share of the even more ambitious task of creating the Panavia Tornado.

As the Panavia aircraft entered service in the late 1970s, along with the European-built F-16s, the Starfighter was phased out. To find surviving fleets of operational starfighters, one must look to the Mediterranean, where the F-104 forms the largest component of the Turkish Air Force. More than 150 of these are F-104G or the Canadian CF-104 equivalent which trades its internal gun for additional fuel. (Canada no longer operates the Starfighter, having handed over its last batch of CF-104 to Turkey in 1986.) A force of 26 RF-104G is available for reconnaissance, while the training role is assigned to five two-seaters – a mixture of TF-104G and ex-Canadian CF-104D. On the opposite side of the Aegean, 60 F-104G and five TF-104G serve with the Greek Air Force, but the Turks have one advantage over their Greek counterparts – a small force of 36 F-104S purchased from Italy.

Italy has only six F-104G and 18 RF-109G (expected to retire in a few years time) plus

18 TF-104G trainers. Most of its Starfighter fleet is the F-104S model, which flew for the first time in 1968. Developed by Aeritalia, this has an improved NASARR R21G/H radar, an uprated GE J79–19 engine, and can carry Sparrow or Aspide radar-guided missiles. The 140 or so F-104S now in Italian service are being upgraded under the ASA program adding an improved Fiat R21G/M1 radar, a new fire-control computer, plus better IFF and EW systems. This should keep them effective until 1995.

Northrop's Veteran Clings On

Another veteran of the 1960s whose numbers are shrinking in NATO service is the Northrop F-5. The concept behind this aircraft dates back to a 1954 US Government study of a possible lightweight and inexpensive export fighter able to meet the requirements of European and Asian nations unable to afford front-line US fighters. Follow two years of private-venture development work, Northrop was given a USAF contract to build its N-156T design as a supersonic trainer, a program which was to result in the T-38 Talon.

Construction of the first N-156T started in May 1958, leading to a inaugural flight by the first of three prototypes in July 1959. Working in parallel with the trainer program, the company also developed the single-seat N-156F Freedom Fighter, and one of the three prototypes was built to the latter standard. The version which would become the F-5 was developed to meet specification SOR 199. In October 1962 the US Defense Department awarded Northrop a $20-million fixed-price contract to build the first F-5s. Maiden flight of the first prototype was in May 1963, and was followed in October of that year by the first production aircraft. Flight testing went smoothly, with the aircraft

Most NATO F-104 Starfighter operators built their own aircraft under licence. Canada built aircraft for its own air arm. When obsolescent, many were passed over to Turkey.

A closely-related Northrop pair, an F-5B (foreground) and a T-38 Talon.

clocking up 1.75 million accident-free air miles during the first year of operation.

Initial Defense Department orders were for around 170 aircraft due to be shipped to US allies under the Military Assistance Program (MAP). NATO nations included under this scheme were Greece and Turkey. Some of these early production aircraft were also passed to the USAF for training purposes and later shipped to Vietnam. By the time that production ended in March 1972, 621 aircraft had been built at a unit flyaway cost of $756,000.

Canada and Spain both took out licences to build the aircraft. By 1971 Canadair had built 89 CF-5A and 46 CF-5B for the Canadian government, plus 75 NF-5A and 30 NF-5B for the Netherlands. More than half of the aircraft built for Canada were kept in storage, the result of lack of operating funds, some later being sold to Venezuela. A second Canadian production built more new aircraft for Venezuela, plus 20 trainers to replace Canada's obsolescent T-33s. Deliveries ended in January 1974.

Only 53 CF-5 remain in Canadian service, a mix of single- and two-seat models. These are being given limited structural and avionics upgrading to allow then to run on into the 1990s, largely in the training role. The Netherlands' fleet of Canadian-built aircraft have had a successful career. A total of 40 NF-5A and 24 NF-5B still serve with 314 and 316 Sqns. The process of retiring the type out started in 1986 when 315 Sqn, re-equipped with the F-16. Next to lose its F-5s was 313 Sqn, and by 1992 the NF-5 will be phased out, to be replaced by the F-16.

CASA's production run ended in 1971, and involved 36 SF-5A and 34 SF-5B being built for the Spanish Air Force. By 1988, 12 SF-5A were still in service, alongside 26 SF-5B and 13 SRF-5A reconnaissance aircraft. Plans are now in hand to replace these in the early 1990s with a new aircraft provisionally designated 'AX'. Planned in single- and two-seat versions, this is likely to be a derivative of the CASA C.101 Aviojet trainer, perhaps with twin engines.

Greece and Turkey, the NATO nations which pioneered the F-5, still operate the type. Greece has 54 F-5A, eight RF-5A, eight F-5B, and 24 ex-Netherlands NF-5A and -5B. Turkey's fleet consists of around 50 F-5A, 20 RF-5A, and at least 15 F-5B trainers, six of which are ex-USAF aircraft taken in charge in 1986. The only other NATO user is Norway, whose 336 Sqn has 17 F-5A, plus 13 F-5B. These aircraft have been reworked to upgrade the avionics and maintain the type in service into the 1990s.

The F-5A may have proved a useful warplane for Third World use, but it lacked the performance required for NATO service. Its twin J85 turbojets developed only 8,000 lb of thrust in full afterburner, while its avionics fit was austere even by the standards of the F-86 Sabre in the Korean War. By the time the F-5A was available in significant numbers the Soviet Union was exporting the faster MiG-21.

The F-5E 'MiG Simulator'

By 1972, Northrop had started deliveries of the faster and more maneuverable F-5E. Fitted with an Emerson nose radar and uprated J85 engines, this aircraft could match the latest models of MiG-21. It found no NATO buyers other than the USAF, which purchased a small fleet to act as 'MiG simulators' with the US Aggressor training units. More than 90 are still in USAF service, along with four F-5F two-seat trainers. In Europe, eyes were already fixed on the USAF's LWF program, which was clearly destined to be an aircraft which would utterly eclipse the limited performance of the Northrop design.

The veteran General Dynamics F-111 (there is no 'popular' name) continues to serve as TAC's main long-range strike aircraft, and is expected to remain in service into the early years of the next century. Current models used are the F-111D, E and F, with a total of more than 325 in service, plus 42 EF-111A Raven EW aircraft. The latter are F-111A aircraft rebuilt by Grumman to incorporate a highly automated version of the US Navy's ALQ-99 jamming system. In the early 1990s, SACs FB-111A nuclear bombers will be modified to handle conventional weapons, re-designated F-111G, and handed over to TAC.

When the USAF found itself in the mid-1960s with an urgent requirement for a low-cost replacement of elderly close-support aircraft such as the Douglas A-1 Skyraider and North American F-100, the logical starting point was the US Navy's A-7 Corsair II. One problem was that the A-7 was short of thrust. It had started life as the A-7A powered by a single 11,350 lb PW TF30-P-6 turbofan, while the follow-on A-7B used the 12,200 lb TF30-P-8.

Rejuvenated Corsair

For its new A-7D, the USAF chose the more powerful Allison TF41-A-1 turbofan, a 14,500 lb thrust derivative of the Rolls-Royce Spey. It also demanded an integrated navigation/bombing system, a 20 mm GE M61 Vulcan cannon, anti-skid brakes and an internal engine starter. The resulting aircraft, which flew for the first time in September 1968, could carry up to 15,000 lb of ordnance on six underwing and two fuselage stations.

The USN was impressed, and took the USAF aircraft as the basis for its own A-7E. The same cannot be said for the rest of NATO, which mostly chose to ignore this effective but subsonic bomb-carrier. Greece was the only customer to sign for new-build Corsairs, buying a fleet of single-seat A-7H and two-seat TA-7H trainers. More than 50 are still operational, as anyone who has holidayed in Western Crete can testify. Vought's (now LTV) 'Sluf' ('Short Little Ugly Fellow') is a common sight in the Cretan skies.

In the late 1970s, Portugal's urgent need for modern combat aircraft was met by rebuilding

20 ex-USN A-7As, changing the existing engine for the slightly more powerful TF-30-P-408, and adding better avionics such as a weapons aiming computer, INS and HUD. A further 50 were supplied in the early 1980s. These A-7P are still operational, 37 serving with 302 and 304 Sqns, alongside six two-seat TA-7P trainers.

The only other NATO nation seriously to consider the A-7 was France, which studied the aircraft in the late 1960s as an alternative to the proposed maritime version of the Sepecat Jaguar. It was rejected in favour of the Dassault Super Etendard. A modified A-7D known as the A-7G was offered to Switzerland in 1967, which would have used the uprated 15,000 lb thrust

Below: **The A-7K is the USAF two-seat trainer version of the Vought Corsair 2.**

Opposite: A SEPECAT Jaguar from 54 Sqn of the Royal Air Force flies a practice mission with four 1,000 lb (450 kg) bombs plus two 1,200 litre external fuel tanks.

Allison TF41-A-3 engine. Although preferred by the Swiss Air Force, it was rejected in favor of second-hand Hunters.

The A-7 production line stayed open at a low rate into the early 1980s, but has now closed. Final model built was the A-7K, a two-seat version of the A-7D produced for the US Air National Guard. As the F-15 Eagle replaced the F-4 Phantom, the latter re-equipped USAF fighter-bomber units thus allowing the A-7 to be handed down to the ANG.

One recent improvement has been the fitting of Low Altitude Night Attack (LANA) systems to 72 A-7D and eight A-7K. This combines an AAR-49 FLIR with a new Singer-Kearfott navigation/weapons delivery computer, and a GEC HUD, allowing the aircraft to carry out low-level night attack missions in all weathers.

By the mid-1980s, the USAF felt that the subsonic performance of the A-7 was not adequate to meet the threat posed by modern Warsaw Pact AA systems. LTV had looked at more potent versions. It had offered the USN the A-7X, a supersonic variant powered either by a single GE F110 or two F404.

In 1985 LTV suggested that the Air Force fleet be rebuilt to meet the needs of the USAF's Close Air Support/Battlefield Air Interdiction requirement. The existing airframes have around two decades of life left, so an upgrade made sense. The scheme originally proposed was designated Strikefighter, being rejected as too expensive, but most of its key ingredients were retained in a less-ambitious proposal known as A-7F Plus. This will install a more powerful engine, modify the wings to increase maneuverability, but the avionics improvements will be less than had been proposed for the Strikefighter.

Two new sections will be added to the fuse-

Below: France's Jaguar strike aircraft have a simpler avionics suite than their RAF counterparts. Obvious external differences between this aircraft and the 54 Sqn aircraft shown opposite are the absence of a nose window for a laser ranger, and the smaller and simpler EW antenna on the vertical fin.

lage – 29.5 in (75 cm) ahead of the wing, and 18 in (46 cm) in the aft fuselage. The increased length and volume will make space available for new avionics and extra fuel. The engine bay will be modified to accept either the PW F100-200 or GE F110 afterburning turbofan; leading-edge root extensions and augmented flaps will be added to the wings, and the leading and trailing edge flaps will be modified. A LANA system will be fitted to those A-7s not already equipped with it, while new avionics could include a ground collision warning system, a better radar-warning receiver, and HOTAS (Hands On Throttle and Stick) controls of the sort fitted to modern fighters such as the F-18. Maximum ordnance load would rise to 17,380 lb (7,900 kg).

Under a $133 million USAF contract awarded in October 1987, two A-7D were rebuilt as YA-7F prototypes, and flight tested in 1989 with F100-200 engines. It is hoped that around 335 Corsairs – a mixture of A-7D and A-7K – would be reworked to the new build standard. The new engines will be PW F100-PW 100s removed from F-15 Eagles when the latter receive their IPE engines. Price tag for each rebuilt aircraft will be around $6.5 million, about half the projected cost of building all-new A-7Fs.

Deliveries are expected to begin in 1991, allowing the A-7F to serve alongside the planned new close-support aircraft (almost certainly the A-16), with the two types arming TAC's planned nine wings – one fewer than at present. Provision is being made to allow the aircraft to be modified to the full Strikefighter at a later date should the USAF want to do this.

Below: **Jaguar can carry a heavier ordnance load than that delivered by the Second World War B-17 Flying Fortress heavy bomber.**

Jaguar Bridges That Gap

Attrition has sharply reduced the size of the fleets of Sepecat Jaguar strike aircraft operated by France and the UK. France purchased 200 (160 Jaguar A strike aircraft, and 40 Jaguar E trainers). About 120 and 36 remain in service. The UK order and split were almost identical – 203 aircraft made up of 165 Jaguar S strike fighters and 38 Jaguar B trainers. More than 100 remain in service, equipping 6, 54, and 41 Sqns and 226 (OCU) in the UK, also 2 Sqn in West Germany.

In specifying equipment fit, the two nations took very different approaches. France was looking for a low-cost replacement for its Super Mysteres and Super Sabres, so kept avionics as simple as possible, while the UK jumped into the world of digital nav/attack systems. France's conservatism paid off in producing a reliable and practical aircraft, but export customers have followed the RAF lead in opting for a full-specification avionics fit.

Jaguar was not a great success on the export market, coming at a time when NATO nations already had their eyes on a new generation of Mach 2 multi-role warplanes such as Tornado and the F-16, while the needs of many non-NATO nations were being met by the Mirage series. Ecuador, Oman and Nigeria all ordered

The centerline store on this RAF Jaguar may look like an external tank, but is in fact a pod filled with reconnaissance sensors. For two decades, such pods (fitted to Jaguars or Phantoms) provided the RAF with a tactical recce capability.

CM
XZ384

small fleets, the only major order being from India, where the type is built under license by HAL.

Britain and France developed their own avionics installations for Jaguar. The French opted for a relatively simple system, while the RAF installed a more sophisticated system which has generally been adopted by export customers. In the RAF, Jaguar played a major role in bridging the gap between simple strike types such as the Canberra and the later Tornado, but the UK's long-term plans were firmly focussed on the latter aircraft. No full mid-life update is planned for the RAF fleet, but aircraft have received the Ferranti FIN 1064 digital nav/attack system in place of much of its original avionics fit.

Bill Gunston once described Jaguar as having the wingspan of a Spitfire, and the range/payload performance of a Lancaster. Its top speed of Mach 1.6 counts for little at the low altitudes at which it was designed to fly and fight. Here is cruises effortlessly at high subsonic speed, carrying up to 10,000 lb of ordnance. Clean, it can prove a tricky opponent, its low-level top speed of Mach 1.1 allows it to outrun most models of MiG-21. During 'Red Flag' training missions in the US, highly trained 'Aggressor' pilots flying the nimble F-5E have learned that in air combat Jaguar can bite back!

Bombs old and new – the RAF Jaguar in the foreground carries conventional 1,000 lb (450 kg) high-explosive bombs, but its companion totes a pair of highly-accurate Paveway laser-guided bombs. The long nose carries a seeker head and steering fins. The latter aircraft also carries a jamming pod under the port wing and a chaff-dispenser under the starboard wing.

The Tornado F.2 was the first interceptor version of the Tornado to enter RAF service. It was rapidly replaced on the production line by the definitive F.3. Surviving F.2s are being rebuilt to the interim F.2A standard.

Tornado Blows In

By the late 1960s, the air forces of NATO could see a need to replace many of their existing tactical aircraft. Belgium, Canada, West Germany, the Netherlands, and Italy were looking towards a replacement for their F-104G Starfighters. The UK had hoped to replace its Vulcans, Canberras, and Buccaneers with a new supersonic, variable-geometry design which is hoped to develop in collaboration with France. This project had collapsed in 1967 when the French pulled out. It too joined the group which was looking at a pan-European collaborative venture.

Starting point for initial discussions in 1968 was a German design known as the *Neue Kampfflugzeug* ('New Battle Aircraft'). Realizing that the end product would be too expensive for their budgets and did not meet their detailed requirements, Belgium and Canada pulled out, followed later by the Netherlands. The other potential partners were able to agree a common design, and in March 1969 formed the Panavia industrial consortium to handle development of the aircraft.

West Germany would have liked to see a US engine used, but in September of that year agreed to a new all-European RB.199 design to be developed by a group known as Turbo-Union. Another consortium known as Avionics Systems Engineeering GmbH was set up to handle the avionics, but this proved unworkable and the job was given to the UK company EASAMS (Elliott-Automation Space and Advanced Military Systems) which subcontracted major shares of the work to national consortia in the other member states – EGS in West Germany and SIA in Italy.

At this point, the aircraft had no name, but was referred to as the Multi Role Combat Aircraft (MRCA). It would later be christened Tornado and by US standards the undertaking was protracted. The year 1969 had also seen the start-up of the F-15 Eagle program, but by the time that the first prototype Tornado flew in 1972, Eagle was entering squadron service. On a US program, it takes two or three years to make the transition from first flight to squadron service. Tornado took six years!

IDS deliveries to the Trinational Tornado Training Establishment (TTTE) at Cottesmore in England started in July 1980. Full-scale production at all three national assembly lines was attained in 1981, but economic problems in the UK and West Germany forced a reduction in the planned production rate, stretching the production program by several years.

Luckily for NATO, the end product was well worth waiting for. Although not much larger than the F-16, Tornado can fly at heights as low as 200 ft (60 m), and carry up to 18,000 lb

MFG 1 (West Germany)

AMI 6 Stormo (Italy)

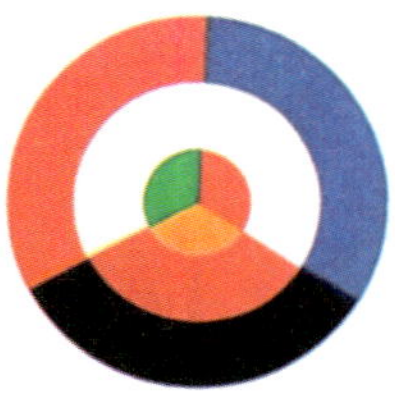

Panavia trinational roundel

(8,180 kg) of ordnance, including nuclear, high-explosive or cluster bombs, Kormoran or Sea Eagle anti-ship missiles, Aspide and Sidewinder air-to-air missiles, or exotic weapons such as Germany's MW-1 submunitions dispenser. It also has two 27 mm Mauser cannons.

Tornado is probably the most versatile multirole aircraft since the 1940s de Havilland Mosquito and Junkers Ju-88. Like these famous predecessors, it also makes a potent interceptor. Italy uses the standard version – known as the Tornado Interdictor/Strike (IDS), but the UK decided in 1976 to develop a specialized fighter version known as the Air Defence Variant (ADV). This has approximately 80 per cent commonality with the IDS version, but has a fuselage stretch of about 4 ft (1.2 m) to make room for its Foxhunter radar and additional fuel.

Development problems with the Foxhunter radar took a long time to sort out, and some ADVs were still flying without radars in 1988. Early ADV aircraft were built to the F.2 standard, and were powered by RB.199 Mk103 engines. The definitive F.3 has Mk104 engines, automatic wing sweep and automatic slat/flap scheduling. All F.3 improvements, apart from the Mk104 engines, will be retrofitted to the F.2s, which will be redesignated F.2A.

Tornado is also used as a reconnaissance aircraft. The Germany Navy and Italian Air Force handle this task by fitting the aircraft with a sensor pod developed by MBB. Britain's RAF wanted a more sophisticated solution, so have built some Tornado IDS with an internally mounted sensor system. These modified Tornado are the first in NATO to have all-electronic reconnaissance sensors. The aircraft's downward-looking IR linescanner and two sideways-looking thermal imagers were first tested in the mid-1980s, while the USAF's equivalent systems are not to be fielded until the early 1990s.

Work on an electronic combat and reconnaissance (ECR) version started in 1984 to meet a West German Air Force requirement. The resulting aircraft is based on the standard IDS variant but carries two AGM-88A Harm missiles, two AIM-9L Sidewinders for self-defense, a jamming pod, a chaff/flare dispenser pod, plus two external fuel tanks. The two 27 mm cannon have been deleted to make room for the internally mounted components of the aircraft's EW system.

The first ECR prototype was IDS development aircraft P.16, which had been rebuilt to carry some items of ECR mission hardware and was followed by a reworked ex-Luftwaffe aircraft. Tests carried out with these aircraft were intended clear the way for delivery in the second half of 1989 of the first production ECR to the Luftwaffe's JBG38 at Jever and JBG 32 at Lechfeld. A fleet of 35 are on order, and all should be handed over by the end of 1991. Panavia has also studied a Wild-Weasel style

'E-Tornado', and offered a similar aircraft to the USAF in 1988 as a candidate airframe for the Follow-On Wild Weasel (FOWW) program.

Tornado IDS already equips nine RAF squadrons, 27 and 617 in the UK, with 9, 14, 15, 16, 17, 20, 31 in West Germany. Deliveries of the type, known to the RAF as the Tornado GR.1, are continuing. West German Air Force orders total 228 IDS and 35 ECR, and the IDS total is likely to be increased to compensate for cancellation of the planned Alpha Jet mid-life update program. Navy orders total 112. The Italian assembly line closed in 1986 following the delivery of 100 IDS.

The Tornado F.2/F.3 (ADV) is not used by Germany or Italy, but already equips two RAF squadrons in the UK – 5 and 29 – plus 229 Sqn, the OCU. Delivery of these first aircraft allowed the veteran BAC Lighting to be phased out early in 1988.

The original three Panavia partners may be delighted with their new aircraft, but no other NATO nation has adopted the type. It was evaluated by Canada in the mid-1980s, but never made that nation's final shortlist. A USAF evaluation in the later 1980s probably never really stood a chance against the F-15E, while Spain and Greece were also to reject it in the early 1980s. The aircraft has since been sold to Jordan, Saudi Arabia, Malaysia and Oman. Turkey has been seen as a possible customer for 40 aircraft, but financing the proposed deal has proved a problem.

Tornado GR.1 is now the RAF's most important strike type. Other examples of what Panavia terms the 'Tornado IDS' also serve with the West German Air Force and Navy, and with the Italian Air Force.

Close Support, Unglamourous but Vital

The most unglamorous tactical mission which any air arm faces is that of close support, the job of backing up the troops on the ground with timely and effective aerial fire support. Three very different types handle this mission on the NATO central front, and all can only be described as controversial.

The USAF's fleet of more than 600 Fairchild A-10 Thunderbolt II make up a major component of the NATO tactical air power available

Panavia Tornado

Role: multi-role combat aircraft
Length: 54 ft 10.25 in (16.72 m)
Height: 19 ft 6.25 in (5.95 m)
Wingspan: 28 ft 2.5 in to 45 ft 7.5 in (8.60–13.91 m)
Weights – empty: 31,065 lb (14,091 kg) **Loaded:** 45,000 lb (20,410 kg) **Max. takeoff:** c.60,000 lb (27,200 kg)
Powerplant(s): two Turbo-Union RB.199 Mk.103 turbofans
Rating: 9,656 lb (4,380 kg) dry thrust, 16,920 lb (7,675 kg) with after burner
Tactical radius: 750 nm (1,390 km) hi-lo-hi with heavy weapons load
Max. range: c.2,100 nm (3,890 km)
Max. speed: Mach 2.2 clean, Mach 0.92 with external stores
Ceiling: 60,700 ft (18,500 m)
Armament: 18,000 lb (8,165 kg) of ordnance plus two 27 mm Mauser cannon

The 'B' designation on the tailfin of this Tornado GR.1 marks it as belonging to the UK component of the Tornado Trinational Training Establishment at RAF Cottesmore in the UK.

Opposite: the striped probe at the nose of this Fairchild A-10 indicates that this is a trials aircraft, probably one of the prototypes.

for the close-support role. The A-10 was developed to meet the demands of the USAF's A-X requirement. Drawn up in the late 1960s, this called for a simple, effective and survivable replacement for the veteran Douglas A-1 Skyraider.

In the past, close support has normally been assigned to obsolete fighters, saving air forces the cost of developing dedicated aircraft for the role, and allowing ageing fighters to perform a useful mission in their twilight years. With the A-10, the USAF started with a clear sheet of paper. The new aircraft was to be able to operate out of unprepared airstrips, loitering in the battle area with heavy ordnance loads, ready to provide firepower and short notice. Such an aircraft would have been the ideal workhorse for the Vietnam War, but its effectiveness in the NATO environment remains, in this author's view, highly questionable. No one would have dreamt of committing the piston-engined A-1 to the Central front in the late 1960s, yet two decades later its jet-powered replacement is tasked with exactly that role. A 400-kt aircraft the size of a World War II medium bomber is being expected to jink and maneuver at treetop height in the face of the most powerful air defenses in history.

To help it survive in combat, it designers went to great lengths to provide a high level of redundancy and to armour all critical areas against most hits by AA rounds of up to 23 mm caliber. It was an ambitious goal, and one which was successfully met. Wing and tail surfaces each have three spars, any two of which can carry the flight loads, the pilot is seated in an armored 'bathtub', while the control runs and fuel lines are protected and well separated. All the fuel is carried in self-sealing tanks, rather than the integral tanks used on most jet fighters. These are filled with reticulate foam and separated by rigid foam spacers as an anti-fire precaution.

Everything that reasonably could be done to ensure survivability was done. No aircraft can be made totally resistant to damage, and a direct hit from a missile or heavy AA round will down to A-10 as surely as any lesser aircraft. In battle, however, many aircraft are lost following relatively minor hits – one rifle bullet in the right place can down an F-111. The A-10A is able to fly home after what Vietnam War aircrew termed the 'golden BB' – the minor hit which can knock out vital controls or systems on a conventional aircraft.

The aircraft's other item of advanced technology is the massive seven-barrelled GAU-8/A internal cannon. This massive 30 mm caliber Gatling-type weapon was developed in parallel with the aircraft, and can pump out 70 rounds per second, each round leaving the barrel at a speed of Mach 3. When the gun is fired its recoil force is approximately equal to that of one of the aircraft's twin TF34 turbofan engines, and ten two-second bursts can empty the aircraft's 1,350-round ammunition maga-

RESCUE

zine. The depleted uranium dart projectile can carve its way through the side or top armour of a tank.

All this survivability and firepower had to be paid for. Target cost of the A-10 was $1.4 million in '1970 dollars', and in the early 1970s aviation author Bill Gunston bet 'a crisp five-pound note' that the unit cost of the aircraft would far exceed the target. His money was safe – a slow down in the production rate took the unit cost to $1.8 million, and changes to the avionics took it all the way to $2.0 million in '1970 dollars'.

Having only the most limited avionics, the A-10A must fly and fight in daylight, and in good weather, the very conditions under which it would be most vulnerable. In the late 1970s, Fairchild did draw up plans for a specialized night attack/all weather version. This was a two-seater based on a FAC-X variant proposed to the USAF in 1976 as a forward air-control aircraft. The night/all-weather (N/AW) version would have had 94 per cent structural commonality with the A-10A and use more than 80 per cent of the same equipment. The major external difference was that the height of the twin vertical tails was increased by around 20 in (51 cm).

Tests carried out with a company-developed demonstrator aircraft showed that the proposed avionics fit – a Texas Instruments AAR-42 FLIR, Westinghouse WX-50 radar, Ferranti laser rangefinder, and a Litton LN-39 inertial navigation system – would allow the aircraft to operate in visibility of less than a mile (1.6 km), and under cloud ceilings as low as 300 ft (90 m),

Below: This A-10A of the 23rd Tactical Fighter Wing is the squadron commander's aircraft. The wheels protrude slightly from the wells so as to be usable during a 'wheels-up' emergency landing.

engaging moving armored vehicles under realistic tactical conditions. Around 300 hours of flight testing, which ended in the spring of 1980, failed to convince the USAF, which anticipated obtaining similar results by equipping the single-seater with the LANTIRN nav/attack pod then in the early stages of development. In practise, the use of LANTIRN is restricted to the F-16 and F-15E, so to this day the A-10A remains an austerely equipped single-seater.

In some respects, the A-10A seems like a rerun of a rather unsuccessful fragment of aviation history. During World War II, the Luftwaffe had its own equivalent of the A-10A in the heavily armored Henschel Hs 129. Like today's A-10, the He 129 was ruggedly built, highly maneuverable, had (by contemporary standards) a modest top speed, and carried a massive gun. On the Eastern front it proved effective, but was never committed to the Western front where the Allies enjoyed air superiority. Without the air superiority needed to keep fighters off its back the He 129 proved highly vulnerable, as would the A-10A at present. In the Thunderbolt II, the USAF built not a re-born P-47, but a re-born Ju 87 Stuka.

This author believes that the decision to deploy the A-10A on a large scale was a mistake, and points to the aircraft's total failure to attract any export orders as evidence of the fact that other air arms also saw the type as an over-specialized and potentially vulnerable jet-age Stuka. The only other nation to embrace the concept was the Soviet Union, whose 1983 deployment of the armored Su-25 'Frogfoot' marked a return to the *Shturmovik* concept of

Below: You can easily see the AGM-65D Maverick missiles for the underwing armament of this A-10A from the New York Air National Guard's 174th Tactical Fighter Wing.

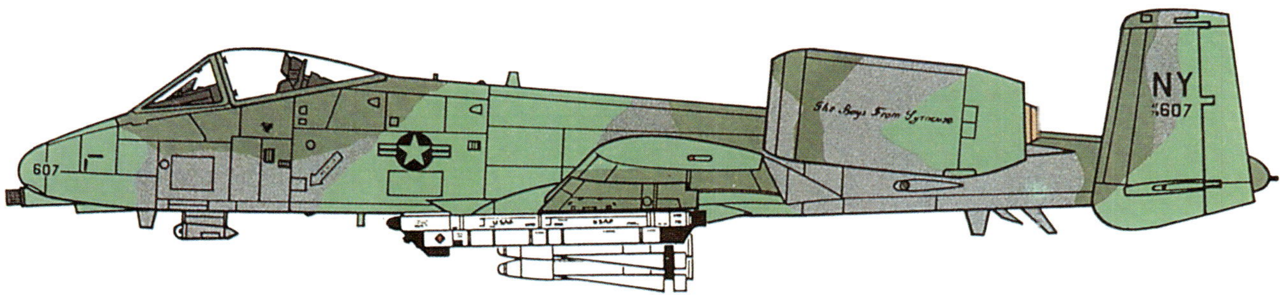

The A-10 does not rely simply on maneuverability and armor protection for its combat survivability. The Pave Penny laser target seeker on the starboard side of the fuselage of this 81st TFW aircraft will speed target acquisition, while the ALQ-119 jamming pod under the starboard wing will counter radar threats.

Fairchild A-10A Thunderbolt II

Role: close-support aircraft
Length: 53 ft 4 in (16.26 m)
Height: 14 ft 8 in (4.47 m)
Wingspan: 57 ft 6 in (17.6 m)
Weights – empty: 21,541 lb (9,771 kg) **Loaded:** 31,831 lb (14,438 kg) **Max. takeoff:** 50,000 lb (22,680 kg)
Powerplant(s): two General Electric TF34-GE-100 turbofans
Rating: 9,065 lb (4,11 kg)
Tactical radius: 540 nm (1,000 km)
Max. range: 2,131 nm (3,949 km)
Max. speed: 450 kts (834 km/hr)
Armament: one GAU-8/A 30 mm cannon plus 14,340 lb (6,500 kg) of ordnance

Opposite: France regards the Dassault/Dornier Alpha Jet simply as an advanced trainer, and an light strike type suitable for Third World use, and has marketed it aggressively for both roles. Despite the sophisticated threat it faces from East European air arms, West Germany regards this graceful little warplane as a front-line combat type.

the 1940s Il-2. But the Soviet Air Force has no dreams of deploying the Su-25 on the scale of the Il-2 and five years of production only 400 aircraft have been built, more than a third of which were for export.

The A-10 relies on high maneuverability at low speed, with pilots making eyeball contact with the target, normally located close to the front line. It is often a tank or other armored vehicle. Two factors have made this aircraft increasingly vulnerable – its low speed, and the growing need to tackle targets such as artillery, which are located much deeper behind enemy lines. All the signs point to the A-10A being retired from 1993 onwards. Some will be reworked as forward air control aircraft, the role for which it was rejected in the late 1970s.

Trainer Turns Into Striker

Probably the oddest aircraft-procurement decision taken during the early 1970s was the Luftwaffe's choice of its next close-support aircraft. Despite being faced by an East German Air Force almost completely equipped with supersonic designs and Soviet air units equipped with the most modern aircraft which Russian industry could create, the Luftwaffe adopted a strike version of a jet trainer – the Dassault-Breguet/Dornier Alpha Jet.

France has no illusions of 'defense on the cheap', so its 164 Alpha Jet E are used for advanced training, as are the 31 of the only other NATO customer, the Belgian Air Force. In an emergency, they will undoubtedly be used in the light strike role as a supplement to more effective types.

The Luftwaffe's 171-strong fleet of Alpha Jet A equips JBG41, 43, and 49, while 18 serve with a weapons-training unit stationed at Beja in Portugal. Originally seen as close support aircraft, the Luftwaffe aircraft are also cleared for airfield defense and the anti-helicopter role.

To boost the Alpha Jet's effectiveness, the Luftwaffe had planned an Improved Combat Efficiency (ICE) program, but this was cancelled in 1987, and in 1989 the aircraft will be given a more modest upgrade. It will be re-engined with the uprated Larzac 04-C20 turbofan fitted with AIM-9 Sidewinders and a clip-on gun pod, and given some limited avionics improvements. Such an equipment fit falls far below what is being offered on the latest Alpha Jet variants. The Lancier version announced in 1985 (also known as Alpha Jet 3) has a Thomson-CSF Agave radar similar to that in the Super Etendard, active and passive EW system, and a FLIR.

There is a place on the military scene for an armed 7,00 kg light trainer with a top speed of 620 mph (1,000 km/hr). The good export sales record of the Alpha Jet bears this out. But in my opinion that place is most definitely not on NATO's central front.

AV-8B/Harrier GR.5 is the latest member of the Harrier family. Visible under the fuselage are the lift-enhancing strakes and flap. These have also been retrofitted USMC A-8A aircraft reworked to the AV-8C standard.

The Logic of V/STOL

When 'Flight International' defense editor Mike Gaines first flew a Harrier sortie in the early 1980s he found the experience an odd one. During the landing phase, just at the time when a pilot should be keeping a close watch on his speed in order to avoid stalling and crashing, the sight of the landscape slowly drifting to a halt was slightly disturbing. By the time he flew the AV-8B in 1988, he found the idea of stopping then landing, rather than landing then stopping, a logical one. This conversion to the joys of vertical operations is not one which is shared by most air arms. Britain's Royal Air Force is the only force in the world able to keep flying tactical sorties with all its runways knocked out. All other users of the AV-8/Harrier family have adopted the type for shipboard or amphibious use.

The British Aerospace Harrier – currently deployed in its GR.3 form – first entered service with the RAF in 1969 as the Harrier GR. 1, becoming the GR.3 in the mid-1970s with the installation of a Ferranti laser ranger and marked target seeker housed in a reprofiled nose. Around 80 remain in service with 3 and 4 Sqns in West Germany, and with 1417 flight in Belize. A mid-life update of the GR.3 had been planned, but this was finally abandoned in favor of buying a larger number of follow-on AV-8B.

The USMC evaluated Harrier in the late 1960s, liked what they saw, and were even prepared to do without a couple of F-4 Phantom squadrons in order to get the Harrier, which was given the US designation AV-8A. They ordered an eventual total of 110, and in the early years of operations did much to explore the concept of 'viffing' – thrust vectoring while in forward flight. Swinging the nozzles of the engine while in air combat allows the Harrier family to fly maneuvers which no conventional fighter can follow, making the V/STOL aircraft a tricky adversary. In the 1980s, 'viffing' is an accepted operating technique – in the early 1970s it was little less than revolutionary.

The USMC purchase marked the start of Harrier's shipboard career. The AV-8A was purchased by Spain, and development of the fully navalised Sea Harrier began in the early 1970s, leading to Royal Navy and Indian Navy orders. Despite the Israeli demonstration in the opening hours of the 1967 Six Day war of just how vulnerable air bases can be, the other NATO air forces stolidly ignored the V/STOL concept, convinced that their runways could survive in wartime.

The Initiative Passes to the US

Initial design work on a new version of Harrier started in the late 1960s. By 1973, BAe and

Left: a Royal Air Force Harrier GR.3 is serviced in a dispersed front-line 'hide'. Few other present-day aircraft could operate from such a primitive site.

Right: foul weather proves no impediment to Harrier operations. If a helicopter can operate, so can this tough little strike aircraft.

McDonnell Douglas were working on the AV-16A Advanced Harrier, an aircraft intended to offer twice the payload/range performance of the existing aircraft. Two years later the UK pulled out, and the program was abandoned. While the UK remained content to to keep the existing Harrier GR.3 in production, pressed on with the minimally modified Sea Harrier, and toyed with the idea of creating its own advanced Harrier, the initiative in V/STOL technology passed from Britain to the United States.

In November 1978, McDonnell Douglas test-flew the first of two prototypes of the proposed AV-8B. Designated YAV-8B, these had been created by rebuilding two AV-8As with a new wing of composite construction, which had greater span and less leading-edge sweep. This led to the construction of four pre-production aircraft.

Meanwhile the UK had flirted with what was then known as the 'Big Wing' Harrier, a design intended to satisfy Air Staff Requirement 409, which placed great emphasis on air combat maneuverability/survivability. AV-8B was seen as too orientated towards the USMC's ambitious payload/range requirements, In 1980, the RAF evaluated the McDonnell Douglas aircraft, discovering that the latter met ASR 409 in all respects except turn rate, a parameter which the UK considered essential if the new fighter was to be able to defend itself against enemy air-superiority fighters. This shortcoming was eradicated by the installation of leading edge wing root extensions (LERXs).

In July 1981 the UK agreed to purchase a minimum of 60 aircraft, and to take part in the AV-8B development program, contributing an initial $80 million, plus the estimated $200 million to be spent in developing the planned RAF configuration. This was designated Harrier GR.5, and incorporated a Ferranti moving-map display in the cockpit, a Martin-Baker ejection seat, additional pylons for AIM-9 Sidewinder missiles and some airframe strengthening.

Under the agreement, the UK would set up its own final assembly line, importing the wings, forward fuselage, horizontal tail, and some items of avionics from the US. The UK would manufacture the rear and center fuselage sections the centred line pylon and the aircraft's reaction control system. Fin and rudder assemblies for RAF aircraft are made in the UK, all others in the US. Worksplit on the aircraft was to be 60 per cent for McDonnell Douglas and 40 per cent for BAe (75/25 on all third-country sales), while the engine was tackled 75 per cent to Rolls-Royce and 25 per cent for Pratt & Whitney.

Production deliveries to the USMC started in October 1983, allowing the first squadron to become operational in January 1985. October of that year saw the first flight of the RAF-standard version. Two prototypes of the GR.5 were flown, clearing the way for deliveries to

The extended nose on the RAF Harrier GR.3 houses a Ferranti laser range finder and marked target seeker, equipment not fitted to the original Harrier GR.1 or the AV-8A. Note the underwing Sidewinder missiles, a weapon fitted to some GR.3 to improve their self-defence capability.

begin in July 1987. Initial deliveries were to 233 Sqn, the GR.5 Operational Conversion Unit. First operational unit was 1 Sqn, which formed in the following year and a second batch of 34 aircraft was ordered in April 1988. Deliveries to the third customer, Spain's naval air arm, started in October 1987, the 12 aircraft being designated EAV-8B.

Night Fighting Harrier

The USMC plans several early updates to the aircraft. By the end of 1989, aircraft should be receiving night attack capability with the provision of a FLIR system, pilot night-vision goggles, and a modified HUD, while 1992 should see the installation of a nose-mounted radar, probably the Hughes APG-65 used in the F-18 Hornet. Once this has been done, the

British Aerospace Harrier GR.3

Role: V/STOL close-support fighter
Length: 46 ft 10 in (14.27 m)
Height: 11 ft 11 in (3.6 m)
Wingspan: 25 ft 3 in (7.70 m)
Weights – empty: 13,535 lb (6,140 kg) **Loaded:** c. 19,000 lb (8,600 kg)
Powerplant(s): one Rolls-Royce Pegasus Mk.103 vectored-thrust turbofan
Rating: 21,500 lb (9,750 kg)
Tactical radius: 360 nm (666 km) hi-lo-hi with 4,400 lb (1,995 kg) of ordnance
Max. range: 1,850 nm (3,425 km)
Max. speed: 635 kts (1,176 km/hr) at sea level
Ceiling: 51,200 ft (15,600 m)
Armament: 5,000 lb (2,270 kg) or ordnance, plus two 30 mm Aden cannon

modified aircraft will be designated Harrier II Plus. Other options under study as part of this upgrading process an an uprated engine offering an extra 3,000 lb (1,360 kg) of thrust, and a digital engine control system. The RAF will not take all of these modifications, put does plan to install the night-attack systems, When this is done, the RAF aircraft will be redesignated Harrier GR.7. The RAF has ordered no TAV-8B trainers, but plans to convert crews to the GR.5 using the two-seat Harrier T.4. When given night-attack systems, these aircraft will be re-designated Harrier T.6.

NATO's least-publicised attack aircraft is probably the Aeritalia/Aermacchi/Embrarer AMX. Intended as a low-cost STOL aircraft able to operate out of semi-prepared strips, AMX started life as an Italian project drawn up to meet a 1977 Air Force specification, with Brazil joining in at an early stage. Aeritalia has a 49 per cent share of the work, and builds the center fuselage, fin and radome, while Aermacchi's 21 per cent covers and forward and rear fuselage. Embraer's share amounts to 30 per cent – the wings, flaps, intakes and horizontal stabilizer.

AMX weighs about 23,700 lb (10,750 kg) fully loaded, and is powered by a single Rolls-Royce Spey turbofan. This has no afterburner, since the aircraft is subsonic. The Italian version has a 20 mm M61 Vulcan cannon, while the Brazilian has two 30 mm DEFA 553 cannon. Hard points under the wings and fuselage can carry up to 8,377 lb (3,800 kg) of ordnance.

Six prototypes were built, four in Italy and two in Brazil. Both nations have their own assembly lines, and intend to build at least 317 aircraft between them. Italy will take 187 single-seaters and 51 AMX-T two-seat trainers, with Brazil taking 65 and 14 respectively. First production deliveries to the Italian Air Force took place in 1988.

In Italian Air Force service, AMX is replacing some F-104 Starfighters, plus another NATO veteran – the Aeritalia G.91 fighter. At present some 165 of these light subsonic strike fighters remain in service with the Italian Air Force – a mixture of around 70 G.91T two-seat trainers and 45 G.91R single-seat fighters (both single-engined), plus around 70 examples of the final twin-engined G.91Y attack model.

West Germany's G.91 force has long been phased out, and the only other NATO ally to purchase the type was Portugal, which still operates two squadrons in the Azores (301 and 303), equipped with a mixture of G.91R-2 and R-4 models. Fewer than 50 remain, but in the absence of potential replacements these are being updated with new avionics and AIM-9 Sidewinder missiles. There are also ten G.91T trainers.

Super Sabre – the Tricky 'Hun'

The other end of the Mediterranean also has its share of yesteryear's warplanes. The North American F-100 Super Sabre, a major NATO

Opposite: the 'hemp' camouflage scheme on this early Royal Air Force Harrier GR.5 was later replaced by an all-green finish. This photo gives a good view of the wing leading-edge root extensions (LERX) fitted to improve the aircraft's instantaneous turn rate. They were originally developed for the proposed all-British 'Big Wing Harrier'.

Below: The UK's successful Experimental Aircraft Programme (EAP) creating a single aircraft which demonstrated the technology for the planned European Fighter Aircraft.

Opposite: John Taylor, editor of the prestigious yearbook 'Jane's All the World's Aircraft', once said that the planned European Fighter Aircraft 're-invents the F-16, with refinements'. Is it the aircraft NATO really needs for the 1990s and beyond, or is it simply the best that Western Europe feels it can afford?

combat type in the 1950s, was known to the men who flew it at the 'Hun', a name which summed up its designation and its reputation of being tricky to fly. Last relics of the once-large fleet consist of about 35 F-100D single-seat fighter bombers still used for operational training in the Turkish Air Force, plus a slightly smaller number of two-seat F-100F trainers.

In the 1990s, many of the fighters currently in NATO service will need to be replaced as the Alliance attempts to offset the growing effectiveness of advanced Warsaw Pact aircraft such as the MiG-29 Fulcrum. To maintain NATO's air power into the 1990s and beyond, four new tactical fighters are under development. Britain, West Germany, Spain and Italy have joined forces to create the Eurofighter industrial consortium which plans to develop a canard delta powered by twin turbofans. Like the consortium, this is also known as Eurofighter.

Like the British Aerospace EAP technology demonstrator which flew for the first time on 8 August 1988, the Eurofighter prototypes due to fly in 1991 will be powered by two Turbo-Union RB.199s (the engine used in Tornado), but the production aircraft will be fitted with engines of all-new design. These will be EJ.200.200 turbofans, a twin-shaft turbofan being developed by the Eurojet consortium of Rolls-Royce (UK), MTU (W. Germany), Fiat (Italy) and Sener (Spain).

Under a memorandum of understanding signed in May 1988, Eurofighter will be built by British Aerospace (UK), MBB (W. Germany), Aeritalia (Italy) and CASA (Spain). Production deliveries are due to begin in 1996. Single and two-seat versions are planned; the basic single-seat version will be designed for the air-to-air role, but will have secondary air-to-surface capability, while the two-seater will be for training. The UK has a requirement for more than 250 Eurofighters, while West Germany plans to buy at least 200, Italy needs 200 and Spain plans a fleet of 100.

Rival European project is the Dassault-Breguet Rafale ('Squall'), a slightly smaller and lighter canard delta powered by two of Snecma's new M-88 twin-shaft turbofans. In its basic Rafale D form, this aircraft is due to fly in 1990, entering service six years later in single and two-seat variants intended to replace the Mirage IIIE and Jaguar. Rafale D will be a single-seat naval version intended to replace the F-8 Crusader and Super Etendard aboard France's aircraft carriers.

The Rafale A technology demonstrator, which flew for the first time on July 4, 1986, exceeding Mach 1.3, is about 1,000 lb (450 kg) heavier than the definitive aircraft, and is powered by two GE F404 turbofans. France had hoped that this prototype aircraft might act as the catalyst for other European nations to join the program, particularly Belgium, which has already license-built the Mirage 5. Belgium

Above: France wants a lighter aircraft than EFA, so is pressing ahead with the Rafale programme. This view of the first prototype shows the aircraft's novel blending of fuselage and inlets, a configuration which reduces radar cross section.

was offered a share of the Rafale program, on the understanding that if it did not buy the aircraft for its own use it would refrain from ordering any rival type for as long as the Rafale remained in production. Belgium refused such a deal, leaving France with the prospect of having to foot the Rafale development bill alone.

'Abyss for Billions'

The project is already under attack as a potential 'Abyss for billions', and fresh overtures to the Eurofighter consortium were reported in the second half of 1988. Despite this, French national prestige and its traditional desire to maintain a strong and independent aerospace industry remain powerful incentives for going it alone.

Much less is known about the two other new fighters – the Lockheed YF-22A and the Northrop YF-23A. Under development to meet the USAF's Advanced Tactical Fighter (ATF) requirement, these are due to fly late in 1989. Two prototypes of each aircraft are being built, one powered by a pair of Pratt & Whitney YF-119 turbofans, the other by two rival General Electric YF-120.

Following a competitive fly-off, one airframe and one engine will be selected for a further five years of full-scale development. The definitive design is expected to fly in 1993, leading to first production deliveries in 1995 or 1996, and operational deployment a year or so later. Under present plans, the ATF will weight no more than 50,000 lb (22,700 kg), and has a cost ceiling of $35 million per aircraft. ATF will be about the size of the F-15, allowing it to use current patterns of hardened aircraft shelters. The USAF is expected to order 750, with the USN later taking up to 550 for carrier-based duties as an F-14 replacement.

New and demanding standards of maintainability and reliability have also been set. For example, the specification demands that the new aircraft be able to offer twice the sortie rate of the F-15 Eagle, while requiring 50 per cent less maintenance. Extensive use will be made of new technology, including stealth. The aircraft is expected to use the latest integrated avionics and to rely on thrust-vectoring for good in-flight maneuverability and STOL capability. This is a demanding requirement which some observers think will be a bigger step forward than the transition in the late 1930s from biplanes, fixed undercarriages and open cockpits to the fighters of the Spitfire/Mustang/Bf 109 class. If this is right, most of the aircraft described in this book will be doomed to early obsolescence.

New patterns of attack aircraft will also be

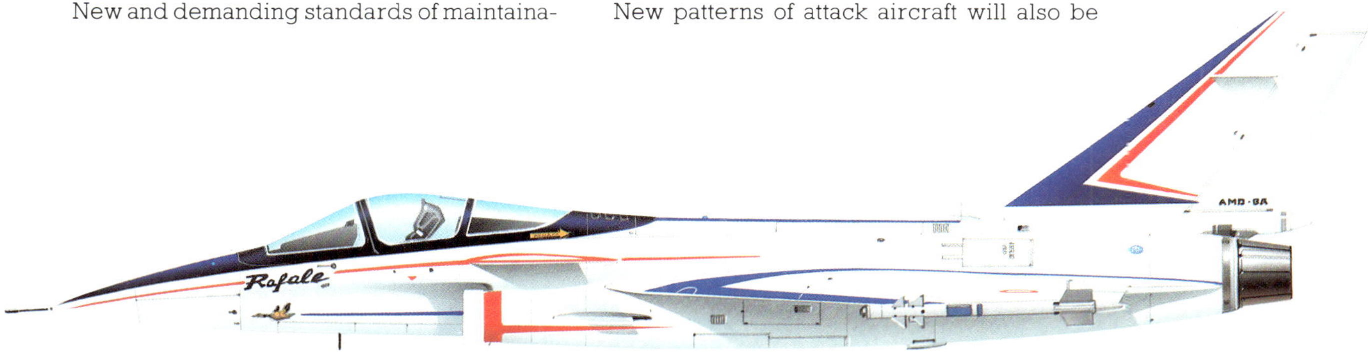

Below: Rafale maintains the Dassault tradition of combining good looks with high performance. On the planned production aircraft, the current General Electric F404 engines will be replaced by French-developed SNECMA M88 turbofans. The cost of developing and deploying this aircraft as an all-French programme will be massive. Like the rival EFA, it could be doomed to early obsolescence by developments in stealth technology.

needed. None is under development in Europe, but two programs are under way in the US. In January 1988, the DoD ordered the USAF to carry out a $20-million study of possible new or modified aircraft able to perform the close-support mission. An all-new aircraft could take up to 11 years to develop, claims the USAF. The estimate seems suspiciously high, even if one ignores critics who argue that the resulting aircraft ought to be small, relatively simple, and built around a 30 mm cannon. Even a more complex design shouldn't take 11 years – the F-16 took only seven years to develop.

It is difficult to avoid the conclusion that the USAF is determined to get approval for a modified version of an existing aircraft, specifically the proposed A-16 derivative of the F-16, and TAC has studied alternatives. A re-engined A-10 would not have enough speed, and would suffer reduced range and loiter time, while the AV-8B, upgraded by the installation of extra avionics and an improved fire-suppression system, was also eliminated as unsatisfactory. TAC considered the cost of RAF-style Harrier operations to be too high. 'Move them [V/STOL fighters] around, and the logistics and communications tail which goes with them becomes very, very, big', TAC commander General Robert D. Russ explained in 1988.

Like the A-10, the A-16 would rely on targeting information from Army airborne and ground controllers, but a secure digital communications link would pass target data to the approaching aircraft, displaying the target coordinates on the pilot's HUD. The USAF hopes that this would allow the target to be knocked out by a single high-speed pass. The technology involved has already been tested on the USAF's experimental AFTI (Advanced Fighter Technology Integration) F-16 aircraft.

The A-16 would have some wing structural improvements including strakes and an automatic maneuvering flap. Other changes include a 30 mm gun pod, a Pathfinder FLIR and a laser spot seeker, a new wide-angle HUD to display the target information, higher capacity chaff and flare dispensers, and a new green-based camouflage scheme for its new role. For the US Army, the unanswered question is whether the A-16s would be available for their intended role in the event of war. Some fear that these high-performance fighters will not go into service alongside the basic F-16 for longer-ranged tasks such as interdiction.

All models of the F-111 will eventually be replaced by a USAF version of the Navy's General Dynamics/McDonnell Douglas A-12 Advanced Tactical Aircraft. Very little is known about this subsonic attack aircraft, development of which started in December 1987. Powered by a pair of F404 turbofans, and incorporating stealth technology, it is expected to enter USN service in the mid-1990s, replacing the Grumman A-6. Like Air Force versions of previous naval aircraft such as the

Asked by journalists at the 1988 Farnborough Air show which Western aircraft at the show most interested them, the pilots who flew Russia's MiG-29 nominated the Harrier. Seen here in its AV-8B form, this aircraft is the only one in NATO's inventory which could keep operating following massive anti-airfield strikes against NATO bases.

An Italian/Brazilian joint venture involving Aeritalia, Aermacchi and Embraer, the Spey-powered AMX is effectively a 'Skyhawk for the 1990s'.

F-4 and A-7, this will be modified to meet the service's requirements. Deliveries to the USAF are expected to begin in 1998.

Future of NATO's Air Arms

The shape of NATO's air arms may change, but their fundamental principle will not. Since the late 1940s, the member states which make up this alliance of former friends and former enemies have remained united in their desire to match the strength of the Warsaw Pact. If the Soviet Government is to be believed, the 1990s will see the Pact slowly change from an offensive to a defensive orientation. Such changes will inevitably be reflected in revised tactics and deployment concepts on this side of the east/west divide, as NATO reacts to the new circumstances. But even if the most optimisitic hopes for disarmament were to be realized, the warplanes described in this book, along with their carrier-based counterparts, will still be needed for many years to come.

Aircraft in Service

This list includes aircraft in the following categories:
Interceptor/Air Superiority, Fighter/Attack, Strike/Attack and Light Attack

Belgium
Dassault-Breguet Mirage VBA
Dassault-Breguet-Dornier Alpha Jet E
General Dynamics F16A/B
Canada
McDonnell-Douglas CF-18/18B
Northrop F5A/D

Denmark
General Dynamics F-16A/B
Saab-Scania F-35

France
Air Force
Dassault-Breguet Mirage F1C, IIIB/E, IVA, VF
Dassault-Breguet Mirage 2000C/N
SEPECAT Jaguar A

Navy
Dassault-Breguet Super Etendard
Vought F-8E

Great Britain
Royal Air Force
British Aerospace Harrier GR.3/GR.5
McDonnell-Douglas Phantom FG.1/FGR.2/F.3
Panavia Tornado GR.1/F.2/F.3
SEPECAT Jaguar GR.1

Fleet Air Arm
British Aerospace Sea Harrier FRS.1

Greece
Dassault-Breguet Mirage F1CG
Dassault-Breguet Mirage 2000E
General Dynamics F-16C
Lockheed F-104G
McDonnell-Douglas F-4E
Northrop F-5A, NF-5A/B
Vought A-7H

Italy
Aeritalia-Aermacchi-Embraer AMX
Fiat G91R/Y
Lockheed F-104S
Panavia Tornado

Netherlands
General Dynamics F-16A/B
Northrop NF-5A/B

Norway
General Dynamics F-16A/B
Northrop F-5A/B

Portugal
Fiat G91R
Vought A-7P

Spain
Air Force
Dassault-Breguet Mirage F1C/E, IIIE
McDonnell-Douglas F-4C
McDonnell-Douglas EF-18A
Northrop SF-5A

Navy
British Aerospace AV-85
British Aerospace EAV-8B

Turkey
General Dynamics F-16C
Lockheed F-104G/S
McDonnell-Douglas F-4E
Northrop F-5A

United States
US Air Force
Fairchild A-10A
General Dynamics F-16A/B/C
General Dynamics F-111D/E/F
Lockheed F-117A
McDonnell-Douglas F-4C/D
McDonnell-Douglas F-15A/C/E
Northrop F-5E/F
Vought A-7D

US Navy
General Dynamics F-16N
Grumman A-6E
Grumman F-14A/D
McDonnell-Douglas A-4E
McDonnell-Douglas F-18A/B/C
Northrop F-5E/F
Vought A-7B/E

US Marine Corps
Grumman A-6E
McDonnell-Douglas A4E/F/M
McDonnell-Douglas F-4N/S
McDonnell-Douglas F-18A/C
McDonnell-Douglas AV-8B

West Germany
Luftwaffe
Lockheed F-104G/TF-104G
McDonnell-Douglas F-4F/RF-4E
Panavia Tornado
Dassault-Breguet-Dornier Alpha Jet

Marineflieger
Panavia Tornado

Acknowledgements

The publishers are grateful to manufacturers for many of the illustrations in this book, to Octopus Books, Pilot Press (pages 8, 17, 20, 21, 22, 23, 24, 27, 28, 43, 45, 48, 70, 72, 73, 74, 75), Quadrant Picture Library (pages 2, 77 and 78) and Jerry Scutts (pages 39, 40 and 53).